AF133613

Advanced Guidance and Control of Flight Vehicle: Theory and Application

Advanced Guidance and Control of Flight Vehicle: Theory and Application

Editors

Haizhao Liang
Jianying Wang
Chuang Liu

Basel • Beijing • Wuhan • Barcelona • Belgrade • Novi Sad • Cluj • Manchester

Editors

Haizhao Liang
Sun Yat-sen University
Shenzhen
China

Jianying Wang
Sun Yat-sen University
Shenzhen
China

Chuang Liu
Northwestern Polytechnical University
Xi'an
China

Editorial Office
MDPI AG
Grosspeteranlage 5
4052 Basel, Switzerland

This is a reprint of articles from the Special Issue published online in the open access journal *Mathematics* (ISSN 2227-7390) (available at: https://www.mdpi.com/si/mathematics/WJK550NMZL).

For citation purposes, cite each article independently as indicated on the article page online and as indicated below:

Lastname, A.A.; Lastname, B.B. Article Title. *Journal Name* **Year**, *Volume Number*, Page Range.

ISBN 978-3-7258-1885-3 (Hbk)
ISBN 978-3-7258-1886-0 (PDF)
doi.org/10.3390/books978-3-7258-1886-0

© 2024 by the authors. Articles in this book are Open Access and distributed under the Creative Commons Attribution (CC BY) license. The book as a whole is distributed by MDPI under the terms and conditions of the Creative Commons Attribution-NonCommercial-NoDerivs (CC BY-NC-ND) license.

Contents

About the Editors vii

Yeguang Wang, Honglin Liu and Kai Liu
Improved Thrust Performance Optimization Method for UAVs Based on the Adaptive Margin Control Approach
Reprinted from: *Mathematics* 2023, 11, 1176, doi:10.3390/math11051176 1

Jing Sheng, Yunhai Geng, Min Li and Baolong Zhu
Finite-Time Contractive Control of Spacecraft Rendezvous System
Reprinted from: *Mathematics* 2023, 11, 1871, doi:10.3390/math11081871 23

Lixia Deng, Huanyu Chen, Xiaoyiqun Zhang and Haiying Liu
Three-Dimensional Path Planning of UAV Based on Improved Particle Swarm Optimization
Reprinted from: *Mathematics* 2023, 11, 1987, doi:10.3390/math11091987 37

Yaosong Long, Chao Ou, Chengjun Shan and Zhongtao Cheng
A Novel Fixed-Time Convergence Guidance Law against Maneuvering Targets
Reprinted from: *Mathematics* 2023, 11, 2090, doi:10.3390/math11092090 50

Chao Ou, Chengjun Shan, Zhongtao Cheng and Yaosong Long
Adaptive Trajectory Tracking Algorithm for The Aerospace Vehicle Based on Improved T-MPSP
Reprinted from: *Mathematics* 2023, 11, 2160, doi:10.3390/math11092160 65

Jiarui Ma, Hongbo Chen, Jinbo Wang and Qiliang Zhang
Real-Time Trajectory Planning for Hypersonic Entry Using Adaptive Non-Uniform Discretization and Convex Optimization
Reprinted from: *Mathematics* 2023, 11, 2754, doi:10.3390/math11122754 80

Zihan Xie, Jialun Pu, Changzhu Wei and Yingzi Guan
Cubature Kalman Filters Model Predictive Static Programming Guidance Method with Impact Time and Angle Constraints Considering Modeling Errors
Reprinted from: *Mathematics* 2023, 11, 2990, doi:10.3390/math11132990 98

Jun Ma, Zeng Wang and Chang Wang
Hybrid Attitude Saturation and Fault-Tolerant Control for Rigid Spacecraft without Unwinding
Reprinted from: *Mathematics* 2023, 11, 3431, doi:10.3390/math11153431 113

Weilin Ni, Jiaqi Liu, Zhi Li, Peng Liu and Haizhao Liang
Cooperative Guidance Strategy for Active Spacecraft Protection from a Homing Interceptor via Deep Reinforcement Learning
Reprinted from: *Mathematics* 2023, 11, 4211, doi:10.3390/math11194211 130

Haizhao Liang, Yunhao Luo, Haohui Che, Jingxian Zhu and Jianying Wang
A Reentry Trajectory Planning Algorithm via Pseudo-Spectral Convexification and Method of Multipliers
Reprinted from: *Mathematics* 2024, 12, 1306, doi:10.3390/math12091306 155

About the Editors

Haizhao Liang

Haizhao Liang, a Full Professor at Sun Yat-sen University, received his Ph.D. degree in Aeronautical and Astronautical Science and Technology from Harbin Institute of Technology (HIT) in June 2013. As a visiting scholar, he worked at Politecnico di Milano between 2011 and 2012. He then worked at the Beijing Institute of Long March Vehicle as an engineer from 2013, and as a senior engineer from 2016. He joined the school of aeronautics and astronautics, Sun Yat-sen University, as an Associate Professor in 2018. His research interests include robust guidance, intelligent guidance, game theory, and active defense strategy.

Jianying Wang

Jianying Wang, an associate professor at Sun Yat-sen University, received her Ph.D. degree in Aeronautical and Astronautical Science and Technology from Harbin Institute of Technology (HIT) in June 2013. As a visiting scholar, she worked at Politecnico di Milano in 2012 with Prof. Franco Bernelli-Zazzera. She then worked at Science and Technology on Space Physics Laboratory as an engineer from 2013, and as a senior engineer from 2016. She joined the school of aeronautics and astronautics, Sun Yat-sen University, as an Associate Professor in 2018. Her research interests include robust control, intelligent control, hypersonic vehicle, trajectory optimization, and trajectory planning.

Chuang Liu

Chuang Liu received his Ph.D. degree in Aeronautical and Astronautical Science and Technology from Harbin Institute of Technology (HIT) in April 2019. As a visiting scholar, he worked at York University between 2017 and 2018. He joined the School of Astronautics, Northwestern Polytechnical University (NWPU) as an Associate Professor in 2019, and is now a Full Professor at NWPU. He was selected for the Young Elite Scientists Sponsorship Program by CAST in 2022. He obtained a COSPAR Outstanding Paper Award for Young Scientists and the first prize of Natural Science Award of Shaanxi Province in 2021. In addition, he serves as an Academic Editor for the 'International Journal of Aerospace Engineering' and peer reviews for more than ten SCI-indexed journals. His research interests include robust controller design, spacecraft dynamics and control, and satellite swarm control.

Article

Improved Thrust Performance Optimization Method for UAVs Based on the Adaptive Margin Control Approach

Yeguang Wang [1,2], Honglin Liu [3,*] and Kai Liu [3]

1 Department of Aeronautics and Astronautics, Fudan University, Shanghai 200433, China
2 Shenyang Aircraft Design and Research Institute, Shenyang 110034, China
3 School of Aeronautics and Astronautics, Dalian University of Technology, Dalian 116024, China
* Correspondence: honglin_liu@mail.dlut.edu.cn

Abstract: This study proposes a strategy for improving the thrust performance of fixed-wing UAV turbine engines from the perspective of aircraft/engine integration. In the UAV engine control process, the inlet distortion caused by the angle of attack change is taken into account, the inlet distortion index is calculated in real time by predicting the angle of attack, and the influence of the inlet distortion on the engine model is analyzed mechanically. Then, the pressure ratio command is adjusted according to the new compressor surge margin requirement caused by the inlet distortion to finally improve the engine thrust performance. To verify the effectiveness of the algorithm, an adaptive disturbance rejection controller is designed for the flight control of a fixed-wing UAV to complete the simulation of horizontal acceleration. The simulation results show that, with this strategy, the UAV turbofan engine can improve the turbofan engine thrust performance by more than 8% under the safety conditions.

Keywords: unmanned aerial vehicle; aircraft/engine integration; thrust optimization; adaptive margin model; adaptive disturbance rejection control

MSC: 37M10

1. Introduction

The unmanned aerial vehicle (UAV), as a rapidly developing aircraft technology, plays an important role in many fields [1–3], such as large-scale display activities, aerial photography entertainment, mapping and geological exploration, agricultural spraying, police patrol, electric power inspection, etc. [4]. UAV propulsion systems have a decisive influence on the performance, cost, and reliability of UAVs. With the continuous enrichment of application scenarios, different types of power units have been developed for different requirements of UAVs in terms of flight speed, flight altitude, maneuvering overload, landing and take-off methods, range, and economic indicators [5]. Most of the power units currently installed on advanced UAVs are turbofan engines, and this status quo will not change for a long time to come. Li conducted work related to the prediction of future UAV power development trends [6], summarized the overloaded advanced typical combat UAVs and the main parameters of power, and compared the payloads of several types of UAVs under development with those of manned fighter aircraft in service. The study shows that the development of large-thrust turbofan engines is one of the future unmanned fighter power development trends. Hu analyzed the requirements of various unmanned aircraft systems for propulsion systems and the impact of propulsion system technology on the performance of UAVs [7,8]. Chen analyzed the development requirements of different types of engines for UAVs and proposed the integrated design of the full authority digital engine control (FADEC) system of a vehicle/engine/propeller for optimal performance matching output [9]. In summary, the research on the improvement of the thrust performance of future UAV turbine engines has considerable value, and this part of the work is not abundant at present. Cheng [10] proposed a turbofan engine Turbine Inter Burner (TIB) supplemental combustion and thrust augmentation thermodynamic cycle scheme based on ultra-compact combustion UCC technology, increasing the engine unit thrust significantly. Dong [11] analyzed the effect of jet pre-cooling on the high- and low-pressure shaft torque, obtained the characteristics of the effect of jet pre-cooling on the control scheme, and also designed the relevant control scheme to adjust the nozzle and buck ratio limiting strategies, thus improving the engine speed performance. Sun [12] established a piecewise linear model of the turbofan engine and used a weighted prediction method to effectively estimate the thrust of the engine and track the set expected thrust to achieve maximum compensation for the thrust loss, and the simulation results showed that the whole thrust control system can ensure the safe operation of the engine and achieve effective thrust boost.

The existing related work lacks the ability to improve powertrain performance from an aircraft/engine integration perspective, and the integrated control of UAVs and engines can seek the best match of their performance. In the aircraft/engine integration research, the stability control technology of the turbo-fan engine is essential. The influence of inlet distortion on the performance and stability of turbine engines is first considered, which is a common phenomenon that occurs widely in turbo-fan engines; the geometric asymmetry of the inlet airflow channel of the compressor, the obstruction of the local inlet airflow by the leading edge of the compressor inlet when the aircraft is flying at a large angle of attack, and the uneven distribution of the temperature and density of the compressor inlet flow field by the missile trailing combustion airflow are the main reasons [13,14]. In order to solve the problem of the unmeasurable engine surge margin in super maneuvering flight, Wang [15] proposed a modeling method of the engine surge margin and verified that the above scheme can accurately control the engine surge margin at 11~13% by the numerical simulation of maneuvering flight with a large angle of attack, which ensures the stability and high efficiency of engine operation. Liu [16] concluded that inlet distortion is an important factor affecting engine stability. To meet the inlet distortion requirements, the engine design needs to make large concessions; in other words, the inlet distortion required for the surge margin greatly reduces the performance of the design point of the compression system. This reduction in margin translates directly into improved system capabilities.

Indeed, the current strategy, which considering flight control and engine control separately, is capable of achieving the optimal performance of each subsystem. However, the overall performance cannot be optimized since the change in engine operation caused by the angle of attack is not taken into account. A large margin of surge is usually given in the control in order to ensure the safety of the compressor operation, which sacrifices a certain engine performance. Inspired by the development of aero-engine stability control technology, this paper proposes a method for improving the thrust performance of future UAVs. This method incorporates engine stability assessment into the engine control process, calculates the requirement of the compressor surge margin associated with the engine stability influence factor in real time, and evaluates the compressor stability, allowing the control system to minimize the surge margin to improve the engine performance and thus enhance the engine thrust performance.

2. Dynamical Modeling and Problem Description

In order to evaluate the UAV thrust performance improvement, UAV flight simulation under certain flight profiles is required, so the UAV dynamics model is first established.

2.1. UAV Dynamics Modeling

2.1.1. UAV Modeling Assumptions

Considering that the mass of the turbine engine-powered UAV is time-varying during flight and its structure is strong and elastic, there are also centrifugal and Gauche accelerations; thus, the aerodynamic forces acting on the UAV are affected by many variables such as its aerodynamic layout, structural elasticity, and flight state, which sometimes cannot be modeled mathematically accurately. Therefore, the primary factors should be considered when modeling the UAV, ignoring the secondary factors. Figure 1 shows a diagram of the coordinate system of the UAV body. The following assumptions need to be considered in the mathematical modeling of scaled-down UAVs.

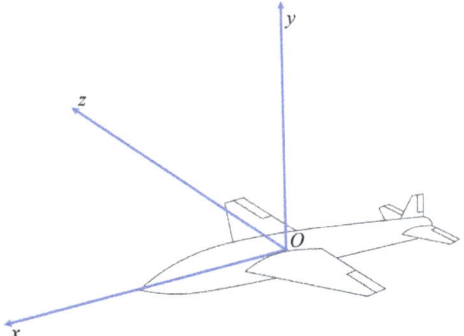

Figure 1. Diagram of the UAV body coordinate system.

- The mass of the UAV does not change during flight for a short period of time.
- The UAV is an ideal rigid body, and the effects of aircraft elasticity are ignored.
- The UAV is affected only by the acceleration of gravity, and the curvature of the Earth is neglected.
- Since the UAV is symmetric about the $x - y$ plane, $I_{xz} = 0$ and $I_{yz} = 0$.
- The change in the aircraft center of mass with fuel is neglected during the UAV.

2.1.2. Dynamics Equations

The dynamics equations for the motion of the center of mass of the UAV are established as follows.

$$\begin{cases} m\dot{V} = P\cos\alpha\cos\beta - X - mg\sin\theta \\ mV\dot{\theta} = P(\sin\alpha\cos\gamma_V + \cos\alpha\sin\beta\sin\gamma_V) + Y\cos\gamma_V - Z\sin\gamma_V - mg\cos\theta \\ -mV\dot{\psi}_V\cos\theta = P(\sin\alpha\sin\gamma_V - \cos\alpha\sin\beta\cos\gamma_V) + Y\sin\gamma_V + Z\cos\gamma_V \end{cases} \quad (1)$$

where m is the UAV mass, V is the UAV velocity, X,Y,Z are the drag lift and lateral force, respectively, P is the UAV thrust, α, β are the angle of attack and sideslip angle, θ is the flight path angle, γ_V is the speed tilt angle, and ψ_V is the trajectory deflection angle.

The dynamics equations of the UAV rotational motion are as follows

$$\begin{aligned} \dot{\omega}_x &= \left(-\frac{(I_{zz}-I_{yy})I_{yy}-I_{xy}^2}{I_{xx}I_{yy}-I_{xy}^2}\omega_y - \frac{(I_{xx}-I_{zz}+I_{yy})I_{xy}}{I_{xx}I_{yy}-I_{xy}^2}\omega_x\right) + \frac{I_{yy}}{I_{xx}I_{yy}-I_{xy}^2}M_x + \frac{I_{xy}}{I_{xx}I_{yy}-I_{xy}^2}M_y \\ \dot{\omega}_y &= \left(\frac{(I_{xx}-I_{zz}+I_{yy})I_{xy}}{I_{xx}I_{yy}-I_{xy}^2}\omega_y - \frac{(I_{xx}-I_{zz})I_{xx}+I_{xy}^2}{I_{xx}I_{yy}-I_{xy}^2}\omega_x\right)\omega_z + \frac{I_{xy}}{I_{xx}I_{yy}-I_{xy}^2}M_x + \frac{I_{xx}}{I_{xx}I_{yy}-I_{xy}^2}M_y \\ \dot{\omega}_z &= -\frac{I_{yy}-I_{xx}}{I_{zz}}\omega_x\omega_y + \frac{I_{xy}}{I_{zz}}\left(\omega_x^2 - \omega_y^2\right) + \frac{M_z}{I_{zz}} \end{aligned} \quad (2)$$

where $\omega_x, \omega_y, \omega_z$ are the angular velocities of rotation of the UAV relative to the ground coordinate system, M_x, M_y, M_z are the components of the moments of all external forces acting on the UAV on the center of mass in the UAV body coordinate system, and $I_{xz}, I_{yy}, I_{zz}, I_{xy}$ are inertia tensor elements.

2.1.3. Kinematic Equations

In the modeling, it is assumed that the UAV is a rigid body and that the six-degrees-of-freedom motion of the UAV is the displacement and rotation around the center of mass.

With the help of the transformation matrix of the ground inertial system and the aircraft dynamic system, the kinematic equations are obtained as follows [17,18].

$$\begin{cases} \dot{x} = V \cos\theta \cos\psi_V \\ \dot{y} = V \sin\theta \\ \dot{z} = -V \cos\theta \sin\psi_V \\ \dot{\vartheta} = \omega_y \sin\gamma + \omega_z \cos\gamma \\ \dot{\psi} = (\omega_y \cos\gamma - \omega_z \sin\gamma)/\cos\vartheta \\ \dot{\gamma} = \omega_x - \tan\vartheta(\omega_y \cos\gamma - \omega_z \sin\gamma) \end{cases} \quad (3)$$

where x, y, z are the center-of-mass position components of the man-machine, ϑ is the pitch angle, and γ is the roll angle.

2.2. UAV Turbofan Engine Modeling

The object of this study is a twin-shaft small-culvert-ratio mixed-exhaust turbofan engine containing no afterburner. The main components of the engine include the inlet, fan, compressor, burner, high-pressure turbine, low-pressure turbine, bypass, mixer, and nozzle; its structural schematic diagram is shown in Figure 2.

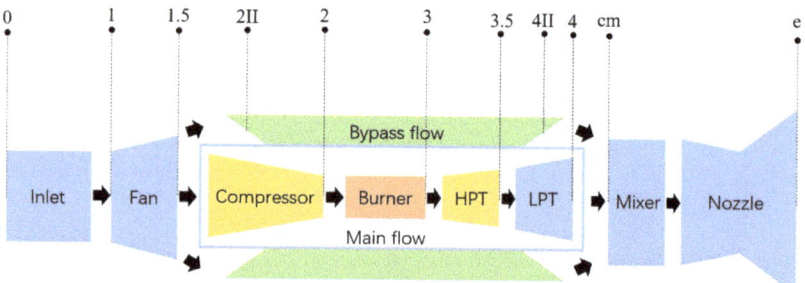

Figure 2. UAV turbofan engine structure schematic.

The aerodynamic and thermal processes during the operation of an aero-engine are very complex, and, when modeling them, the real physical processes need to be suitably simplified, and the following assumptions are usually made when modeling at the component level [19].

- Ignore the effect of the Reynolds number variation on the characteristics of the engine components; ignore the effect of atmospheric humidity on the parameters of the thermodynamic properties of the gas; the effect of rotor inertia is considered.
- The flow of air in the engine is treated as quasi one-dimensional flow, i.e., the airflow parameters are considered to be the same in the same cross-section of the engine.

The idea of component-level modeling is to first establish each component sub-model; then, build a non-linear system of equations based on the aerodynamic thermodynamic laws governing the gas flow process, combined with the assumptions of flow continuity and energy balance; finally, determine the engine operating state by solving the system of common operating equations to construct the engine model.

When a small-culvert-ratio turbofan engine enters into quasi-steady-state operation, each component needs to satisfy a series of common operating conditions in mechanical and aerodynamic aspects, including flow continuity and power balance. It is generally accepted that the inlet and outlet sections of each component meet the flow continuity condition and the two rotors meet the power balance condition, i.e., the power from the high-pressure turbine is balanced with the power consumed by the fan and booster stage, and the power from the low-pressure turbine is balanced with the power consumed by the high-pressure compressor. The dynamic engine simulation process also requires an iterative solution of the common set of operating equations, unlike the steady-state simulation, where the flow continuity balance condition is still satisfied, but the power balance condition is no longer satisfied. The high- and low-pressure rotor speeds are no longer used as initial guess variables but are determined by the speed and acceleration at the previous moment, i.e., the power balance conditions are transformed into the rotor dynamics equations.

2.2.1. Inlet

As the inlet section of the engine, the calculation of the relevant parameters of the air inlet tract is related to the height of the engine. The main parameters of the inlet cross-section—atmospheric pressure P_H in P_a and atmospheric temperature T_H in K at different altitudes (H km)—can be obtained by the atmoscoesa function in MATLAB. The inlet pressure p_1^*, temperature T_1^*, and flight speed c_0 of the engine are calculated as follows.

$$\begin{cases} p_1^* = \sigma_i p_H \left(1 + \frac{k-1}{2} M_H^2\right)^{\frac{k}{k-1}} \\ T_1^* = T_H \left(1 + \frac{k-1}{2} M_H^2\right) \\ c_0 = M_H \times a_H = M_H \sqrt{kRT_H} \end{cases} \quad (4)$$

in the above equation, M_H is the Mach number; σ_i is the total pressure recovery factor, where $\sigma_i = \sigma_i(M_H)$; k is the adiabatic index; and a_H is the local speed of sound.

2.2.2. Fan

Assuming the same main and bypass flow pressure ratios and efficiencies of the fan, we have:

$$\pi_{cL}^* = \pi_{cII}^*, \ \eta_{cL}^* = \eta_{cII}^* \quad (5)$$

where π_{cL}^*, π_{cII}^* represent the pressure ratios of the main and bypass flow and η_{cL}^*, η_{cII}^* represent the overall efficiencies of the main and bypass flow.

The exit section airflow parameters of the turbofan engine fan are calculated as follows.

$$p_{1.5}^* = \pi_{cL}^* p_1^*$$
$$p_2^* = p_{1.5}^*$$
$$m_a = m_{aL,cor} \frac{p_1^*}{101325} \sqrt{\frac{288}{T_1^*}}$$
$$T_{1.5}^* = T_1^* \left(1 + \frac{\pi_{cL}^{*r} - 1}{\eta_{cL}^*}\right) \quad (6)$$
$$T_{2II}^* = T_{1.5}^*$$
$$\eta_{cL}^* = \eta_{cL}^*(n_{cL,cor}, \pi_{cL}^*, \phi_L)$$
$$m_{aL,cor} = m_{aL,cor}(n_{cL,cor}, \pi_{cL}^*, \phi_L)$$

in the above equation, $r = \frac{k-1}{k}$, where $k = k(T_{1.5}^*)$; m_a is the total air flow through the fan; $m_{aL,cor}$ is the converted air flow; ϕ_L is the fan adjustable parameter; η_{cL}^* is the converted speed at different temperatures. The last two equations in Formula (6) show that the specific values of $m_{aL,cor}$ and η_{cL}^* are obtained by interpolating the variables associated with each of them.

2.2.3. Compressor

Similar to the fan characteristics, the generic characteristics of a compressor are calculated as follows.

$$m_{aH,cor} = m_{aH,cor}(n_{cH,cor}, \pi_{cH}^*, \phi_H)$$
$$\eta_{cH}^* = \eta_{cH}^*(n_{cH,cor}, \pi_{cH}^*, \phi_H) \quad (7)$$

where $m_{aH,cor}$ is the converted air flow; ϕ_H is the compressor adjustable parameter; $n_{cH,cor}$ is the converted speed of the compressor.

If the fan control law $\phi_H = \phi_H(n_{cH}^*)$ is known, then the mathematical relationship between the conversion speed and the rotational speed $n_{cH,cor} = n_H\sqrt{288/T_{1.5}^*}$, and we can obtain $n_{cH,cor}$ and π_{cH}^*. Then, from $n_{cH,cor}$, π_{cH}^*, and ϕ_H, we can calculate $m_{aH,cor}$ and η_{cH}^*, where the relevant parameters for the outlet section of the compressor are calculated as follows:

$$p_2^* = \pi_{cH}^* p_{1.5}^*$$
$$m_{a2} = m_{aH,cor} \frac{p_{1.5}^*}{101325}\sqrt{\frac{288}{T_{1.5}^*}} \tag{8}$$
$$T_2^* = T_{1.5}^*\left(1 + \frac{\pi_{cH}^{*r}-1}{\eta_{cH}^*}\right)$$

2.2.4. Burner

The burner characteristics are generally given by the component characteristics test; in general, the burner efficiency η_b and the total pressure recovery coefficient σ_b are the main characteristics parameters of the combustion chamber and are calculated as follows.

$$\eta_b = \eta_b(\tilde{\alpha}, p_2^*, T_2^*, T_3^*)$$
$$\sigma_b = \sigma_b(c_b, \theta)$$
$$\tilde{\alpha} = \frac{m_{a2}}{m_f \cdot 14.8} \tag{9}$$
$$\theta = T_3^*/T_2^*$$

in the above equation, $\tilde{\alpha}$ is the residual gas coefficient of the gas mixture; c_b is the gas flow rate; θ is the heating ratio.

The η_b is calculated from $\tilde{\alpha}, p_2^*, T_2^*$, and T_3^* (the burner characteristics), and then the T_3^* is calculated from the energy balance (by iteration). The energy balance equation is calculated as follows.

$$\frac{m_{a2}}{m_f}\left[H_u \eta_b + h_f(T_{f0}) - h_f(T_3^*)\right] - h_g(T_3^*) + h_a(T_2^*) = 0 \tag{10}$$

in the formula above, T_{f0} is the fuel temperature, and h_a, h_f, and h_g denote the enthalpy per unit (kg) of mass of air, fuel, and fuel after vaporization, respectively.

In summary, the combustion chamber output parameters can be calculated as follows.

$$p_3^* = \sigma_b p_2^*$$
$$m_{g3} = m_{a2} + m_f \tag{11}$$

2.2.5. High-Pressure Turbine (HPT) and Low-Pressure Turbine (LPT)

The efficiencies of the high-pressure turbine η_{TH}^*, the gas flow rate $m_{gH,cor}$, and the conversion speed $n_{TH,cor}$ are related to the expansion ratio π_{TH}^*, i.e., its operating state can be determined from π_{TH}^* and $n_{TH,cor}$. The mathematical representation of $n_{TH,cor}$ and $m_{gH,cor}$, obtained from the characteristic curve diagram, is as follows.

$$m_{gH,cor} = m_{gH,cor}(n_{TH,cor}, \pi_{TH}^*)$$
$$\eta_{TH}^* = \eta_{TH}^*(n_{TH,cor}, \pi_{TH}^*) \tag{12}$$

where $n_{TH,cor} = n_H\sqrt{288/T_3^*}$, and T_3^* is the engine turbine inlet section temperature in the design condition.

In summary, the high-pressure turbine outlet parameters are calculated as follows.

$$m_{g3.5} = m_{gH,cor} \frac{p_3^*}{101325} \sqrt{\frac{288}{T_3^*}}$$

$$p_{3.5}^* = \frac{p_3^*}{\pi_{TH}^*} \quad (13)$$

$$T_{3.5}^* = T_3^*[1 - (1 - \pi_{TH}^{*-r'})\eta_{TH}^*]$$

in the above equation, $r' = \frac{k'-1}{k'}$, where $k' = k'(T_3^*, \tilde{\alpha})$ is the adiabatic coefficient of fuel oil after vaporization.

The structures of the high-pressure turbine and the low-pressure turbine are the same, as are the models, and the parameters of the low-pressure turbine outlet section can be obtained.

$$m_{g4} = m_{gL,cor} \frac{p_{3.5}^*}{101325} \sqrt{\frac{288}{T_{3.5}^*}}$$

$$p_4^* = \frac{p_{3.5}^*}{\pi_{TL}^*} \quad (14)$$

$$T_4^* = T_{3.5}^*[1 - (1 - \pi_{TL}^{*-r'})\eta_{TL}^*]$$

in the above equation, $r' = \frac{k'-1}{k'}$, and $k' = k'(T_{3.5}^*, \tilde{\alpha})$ is the thermal insulation index. $m_{gL,cor}$, π_{TL}^*, and η_{TL}^* correspond to $m_{gH,cor}$, π_{TH}^*, and η_{TH}^*, representing the characteristics of the low-pressure turbine.

2.2.6. Mixer

The gas flow rate contained in the mixing chamber is calculated as follows:

$$m_{g4} = K_g' \frac{p_4^* A_{4I} q(\lambda_4)}{\sqrt{T_4^*}} \quad (15)$$

Equivalently, the above equation is transformed as:

$$q(\lambda_4) = \frac{m_{g4} \sqrt{T_4^*}}{p_4^* A_{4I} K_g'} \quad (16)$$

where $K_g' = \sqrt{\frac{k'}{R'} \left(\frac{2}{k'+1}\right)^{\frac{k'+1}{k'-1}}}$, R' is the gas constant, $k' = k'(T_4^*, \tilde{\alpha})$, and A_{4I} is the cross-sectional area of the inner culvert (main flow) into the mixing room.

Additionally, since $q(\lambda_4)$ can be expressed as

$$q(\lambda_4) = \left(\frac{k'+1}{2}\right)^{\frac{1}{k'-1}} \lambda_4 \left(1 - \frac{k'-1}{k'+1} \lambda_4^2\right)^{\frac{1}{k'-1}} \quad (17)$$

the above equation leads to λ_4, which, in turn, gives

$$\pi(\lambda_4) = \left(1 - \frac{k'-1}{k'+1} \lambda_4^2\right)^{\frac{1}{k'-1}} \quad (18)$$

$$f(\lambda_4) = \left(\lambda_4^2 + 1\right) \left(1 - \frac{k'-1}{k'+1} \lambda_4^2\right)^{\frac{1}{k'-1}} \quad (19)$$

$$p_4 = \pi(\lambda_4) p_4^* \quad (20)$$

Again, the main flow and bypass flow pressures are equal, i.e., $p_{4II} = p_4$, so we have

$$\pi(\lambda_4) = \frac{p_{4II}}{p_{4II}^*} = \frac{p_4}{\sigma_{II} p_{2II}^*} \quad (21)$$

$$\lambda_{4II} = \sqrt{\frac{k+1}{k-1}\left\{1 - [\pi(\lambda_{4II})]^{\frac{1}{k-1}}\right\}} \tag{22}$$

So, there is

$$q(\lambda_{4II}) = \left(\frac{k+1}{2}\right)^{\frac{1}{k-1}} \lambda_{4II} \left(1 - \frac{k-1}{k+1}\lambda_{4II}^2\right)^{\frac{1}{k-1}} \tag{23}$$

$$f(\lambda_{4II}) = \left(\lambda_{4II}^2 + 1\right)\left(1 - \frac{k+1}{k-1}\lambda_{4II}^2\right)^{\frac{1}{k-1}} \tag{24}$$

$$m_{4II} = K_q' \frac{p_{4II}^* A_{4II} q(\lambda_{4II})}{\sqrt{T_{2II}^*}} \tag{25}$$

where A_{4II} indicates the area of the bypass duct at the entrance to the mixing room.

The mixer outlet pressure is calculated as follows.

$$p_{cm}^* = \sigma_{cm} \frac{m_{g4} p_4^* + m_{aII} p_{4II}^*}{m_{g4} + m_{aII}} \tag{26}$$

where σ_{cm} is the total pressure recovery coefficient of the mixer.

λ_{cm} can be calculated from $f(\lambda_{cm})$, and then $\pi(\lambda_{cm})$ and $q(\lambda_{cm})$ can be calculated from the pneumatic function calculation formula. In summary, the gas flow and temperature at the outlet section of the mixing chamber are calculated as follows.

$$m_{gcm} = m_{g4} + m_a - m_{aH}$$
$$T_{cm}^* = \frac{c_p T_{2II}^* m_{aII} + c_p' T_4^* m_{g4}}{c_p'' m_{gcm}} \tag{27}$$

where $c_p = c_p(T_{2II}^*)$, $c_p' = c_p'(\tilde{\alpha}, T_4^*)$, and $c_p'' = c_p''(\tilde{\alpha}, T_{cm}^*)$.

2.2.7. Nozzle

According to the different relationship between the ratio of the total pressure of the nozzle outlet p_e^* and the total atmospheric pressure p_H and the nozzle pressure ratio π_{ecr}, the engine nozzle can be divided into subcritical, critical, and supercritical operating conditions.

When $\frac{p_e^*}{p_H} < \pi_{ecr}$, the nozzle is operating in a sub-critical condition. At this point, the calculation of p_e^* and π_{ecr} are as follows.

$$p_e^* = \sigma_e p_{cm}^* \tag{28}$$

$$\pi_{ecr} = \frac{p_e^*}{p_{cr}} = \left(\frac{k'+1}{2}\right)^{\frac{k'}{k'-1}} \tag{29}$$

At this point, the exit section of the nozzle airflow velocity is less than the speed of sound, so $\lambda_e < 1$. The gas in the nozzle is completely expanded; at this time, the nozzle exit section pressure and the external atmospheric pressure are the same, that is, $p_e = p_H$.

From p_e^* and p_e, we have

$$\pi(\lambda_e) = \frac{p_e^*}{p_e} = \frac{p_e^*}{p_H} \tag{30}$$

from $\pi(\lambda_e)$, we obtain λ_e and find $q(\lambda_e)$.

When $\frac{p_e^*}{p_H} = \pi_{ecr}$, the nozzle works in the critical state; at this time, $\lambda_e = 1$ and $q(\lambda_e) = 1$. At this time, the gas in the nozzle is completely expanded, and the nozzle outlet section pressure and the external atmospheric pressure are the same, that is, $p_e = p_H$.

When $\frac{p_e^*}{p_H} > \pi_{ecr}$, the nozzle operation is in the supercritical state; at this time, $\lambda_e = 1$ and $q(\lambda_e) = 1$. At this time, the gas in the nozzle is not fully expanded, and the nozzle outlet cross-section pressure is greater than the atmospheric pressure, that is, $p_e > p_H$.

Therefore, its operating condition can be determined from the nozzle parameters (λ_e, $q(\lambda_e)$, p_e), and the parameters related to the exit section of the tail nozzle are calculated as follows.

$$m_{ge} = K_q' \frac{p_e^* A_e q(\lambda_e)}{\sqrt{T_{cm}^*}} \tag{31}$$

$$c_e = \varphi_e \lambda_e a_{cr} = \varphi_e \lambda_e \sqrt{\frac{2k'}{k'+1} R' T_{cm}^*} \tag{32}$$

In the above equation, m_{ge} is the nozzle outlet flow rate, c_e is the exhaust velocity, a_{cr} is the critical speed of sound, and φ_e is the nozzle flow loss coefficient number.

2.3. Problem Description

As mentioned in the introduction, the aim of this paper is to improve the thrust performance of the UAV by reducing the surge margin reserve of the turbine engine. The real-time estimation of the surge margin is caused by many factors, including inlet distortion, and calculations of the proper pressure ratio of the fan and the compressor are carried out to satisfy the surge margin requirements through multivariable control.

From a mathematical point of view, the thrust lifting problem under this line of thought can be described as the optimization problem of engine thrust regulation control.

$$\begin{cases} \text{Max} & P_{opt}\left(\pi_{cH}^*, m_{aH,cor}, A_8, H, Ma, N_{fan}\right) \\ \text{s.t.} & SM_{opt} < SM_{re} \\ & T_{3.5}^* < T_{cr} \\ & p_2^* < p_{2,cr}^* \end{cases} \tag{33}$$

where P_{opt} is the current state of the engine output thrust; in this optimization problem, the engine output thrust considered is related to the compressor pressurization ratio π_{cH}^*, the compressor converted air flow rate $m_{aH,cor}$, the fan speed N_{fan}, the ambient altitude H, and the flight Mach number Ma. SM_{opt} and SM_{re} represent the current surge margin and the current allowable surge margin, respectively. $T_{3.5}^*$ and T_{cr} are the turbine front temperature and the threshold of the turbine front temperature protection, p_2^* and $p_{2,cr}^*$ represent the current total compressor outlet pressure and the overpressure protection threshold. SM is described as

$$SM = \left[\frac{\left(\frac{\pi_{cH}^*}{m_{aH,cor}}\right)_s - \left(\frac{\pi_{cH}^*}{m_{aH,cor}}\right)_o}{\left(\frac{\pi_{cH}^*}{m_{aH,cor}}\right)_o} \right]_{N_c=const} \times 100\% \tag{34}$$

The problem is constrained by three inequalities: the first is the surge margin of the engine, that is, the compressor does not surge; the second is, in the optimization process, the turbine front temperature cannot exceed the overtemperature protection temperature while the engine is working; the third inequality constraint is that the total compressor outlet pressure of the engine cannot exceed the limit pressure.

In this optimization problem, the independent variables are π_{cH}^*, $m_{aH,cor}$, H, Ma, and N_{fan}, where H and Ma are environment variables that change with the motion of the vehicle and the surrounding environment. The value of N_{fan} can be determined by π_{cH}^*, so the independent variables that can be adjusted in this optimization problem are π_{cH}^*, $m_{aH, \text{ and } cor}$, and the dependent variable is the thrust P_{opt}.

According to the engine model established in Section 2.2, for a known compressor operating state, we can obtain a set of engine operating parameters. In the engine control, π_{cH}^*, as one of the main parameters controlling the compressor operating catch state, can directly affect the working state of the compressor. The physical principle is to change the deflection angle of the guide vane and increase the work carried out by the compressor on the airflow such that the pressure of the airflow at the exit of the compressor is greater. The gas mixture entering the burner carries more energy, so it can increase the overall reasoning performance; however, if π_{cH}^* is too large, it will lead to a reduction in SM, that is, it will lead to the occurrence of the gas turbulence of the compressor, so another control parameter $m_{aH,cor}$ should be adjusted at the same time to limit the appearance of dangerous situations.

3. Major Studies

3.1. Thrust Optimization Based on the Adaptive Margin Model

In the introduction, the idea of thrust promotion is introduced, which is to reduce the compressor surge margin reserve and increase the propulsion performance under the premise of ensuring the safe operation of the engine; a detailed description of this strategy is provided below.

As shown in Figure 3, the first step is to predict the angle of attack $\hat{\alpha}$ in a short period of time in the future by building a dynamic model of the UAV, which is combined with the current mission; then, the distortion index W_a is determined by the current compressor gas conversion flow $m_{aH,cor}$, and the available pressure ratio increment $\Delta \pi_{cH}$ is determined by the distortion index and the current surge margin value of the engine; then, the final pressure ratio correction plan is given according to the protection limit of the turbine engine, and the instruction correction quantity is transmitted to the controller to achieve the goal of thrust lifting.

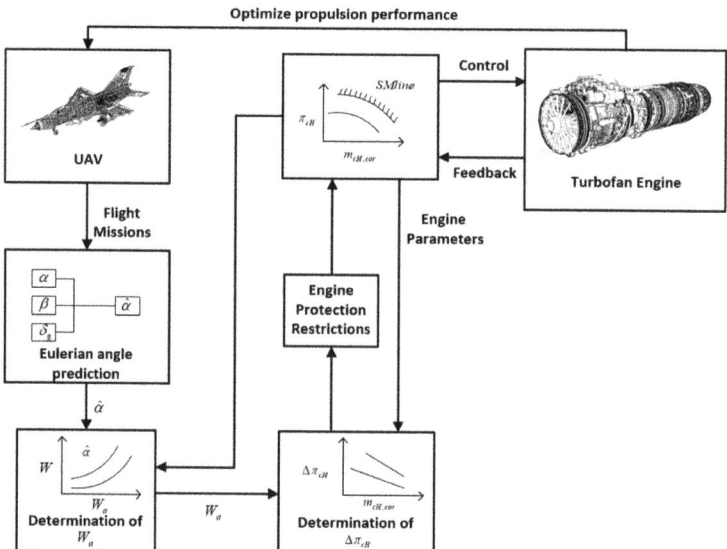

Figure 3. Optimal thrust strategy schematic diagram based on the adaptive margin model.

3.1.1. Angle of Attack Prediction for UAV

Based on the symmetry and small perturbation approximations theory [20], the UAV dynamic model is expanded on the point of longitudinal motion equilibrium and rewritten as perturbation equation. The following equations of motion for the longitudinal disturbance of the UAV are given.

$$\begin{aligned}
\frac{d\Delta V}{dt} - a_{11}\Delta V - a_{13}\Delta \theta - a_{14}\Delta \alpha &= a_6 F_{gx} \\
\frac{d^2\Delta \theta}{dt} - a_{21}\Delta V - a_{22}\frac{d\Delta \theta}{dt} - a_{24}\Delta \alpha - a'_{24}\frac{d\Delta \alpha}{dt} &= a_{25}\Delta \delta_z + a_{26}M_{gx} \\
\frac{d\Delta \theta}{dt} - a_{31}\Delta V - a_{33}\Delta \theta - a_{34}\Delta \alpha &= a_{35}\Delta \delta_z + a_{36}F_{gy} \\
-\Delta \vartheta + \Delta \theta + \Delta \alpha &= 0
\end{aligned} \quad (35)$$

where F_{gx}, F_{gy} denote the disturbance force due to the disturbance introduced during the small disturbance linearization process and the disturbance torque M_{gx}. δ_z denotes the elevator deflection angle. $a_{11} \ldots a_{36}$ are known as the power factors, which characterize the dynamics of the UAV. The specific expressions are shown in Table 1.

Table 1. The symbol of the dynamic coefficient a_{ij} and its expression.

Equations of Motion Number 'i'	Coefficient of Motion Serial Number 'j'					
	1	2	3	4	5	6
1	$a_{11} = \frac{P^V - X^V}{m}$ (s^{-1})	$a_{12} = 0$	$a_{13} = -g\cos\theta$ $(m \cdot s^{-2})$	$a_{14} = -\frac{X^\alpha + P^\alpha}{m}$ $(m \cdot s^{-2})$	-	$a_{16} = \frac{1}{m}$ (Kg^{-1})
2	$a_{21} = \frac{M_z^V}{I_{zz}}$ $(m^{-1} \cdot s^{-1})$	$a_{22} = \frac{M_z^{\omega_z}}{I_{zz}}$ (s^{-1})	$a_{23} = 0$	$a_{24} = \frac{M_z^\alpha}{I_{zz}}$ (s^{-2}) $a'_{24} = \frac{M_z^{\dot\alpha}}{I_{zz}}$ (s^{-1})	$a_{25} = \frac{M_z^{\delta_z}}{I_{zz}}$ (s^{-2}) $a'_{25} = \frac{M_z^{\dot\delta_z}}{I_{zz}}$ (s^{-1})	$a_{26} = \frac{1}{I_{zz}}$ $(Kg^{-1} \cdot m^{-2})$
3	$a_{31} = \frac{P^V\alpha + Y^V}{mV}$ (m^{-1})	$a_{32} = 0$	$a_{33} = \frac{g\sin\theta}{V}$ (s^{-1})	$a_{34} = \frac{P + Y^\alpha}{mV}$ (s^{-1})	$a_{35} = \frac{Y^{\delta_z}}{mV}$ (s^{-1})	$a_{36} = \frac{1}{mV}$ $(s \cdot Kg^{-1} \cdot m^{-1})$

In the table above, $X^\alpha, \cdots, Y^V, \cdots, M_z^{\dot\delta_z}, \cdots, P^V$ represent the partial derivative of parameters such as the aerodynamic torque and thrust with respect to the parameter $\alpha, \cdots, V, \cdots, \dot\delta_z, \cdots, \omega_z$.

In the longitudinal motion of UAV, the outputs of the airframe are $\Delta V, \Delta\vartheta, \Delta\theta$, and $\Delta\alpha$, while the input is $\Delta\delta_z$. If there is external interference, the input is F_{gx}, F_{gy}, and M_{gz}. In order to obtain the transfer function of UAV short-period longitudinal perturbation motion, the long-period output ΔV is first ignored, and the Equation (40) are transformed by Laplace transform as

$$\begin{bmatrix} s(s - a_{22}) & 0 & -(a'_{24}s + a_{24}) \\ 0 & s - a_{33} & -a_{34} \\ -1 & 1 & 1 \end{bmatrix} \begin{bmatrix} \Delta\vartheta(s) \\ \Delta\theta(s) \\ \Delta\alpha(s) \end{bmatrix} = \begin{bmatrix} a_{25} \\ a_{35} \\ 0 \end{bmatrix} \Delta\delta_z(s) + \begin{bmatrix} a_{26}M_{gz}(s) \\ a_{36}F_{gy}(s) \\ 0 \end{bmatrix} \quad (36)$$

Then, the transfer function from the elevator to the angle of attack is obtained as

$$W_{\delta_z}^\alpha(s) = \frac{\Delta\alpha(s)}{\Delta\delta_z(s)} = \frac{-a_{35}s^2 + (a_{35}a_{22} + a_{25})s}{s[s^2 + (a_{34} - a_{22} - a'_{24})s - (a_{34}a_{22} + a_{24})]} \quad (37)$$

So far, we have obtained the transfer function from the elevator deflection command to the change in the angle of attack in the short-period longitudinal motion of the UAV, by which the current flight parameters of the UAV and the elevator command can be combined, and the angle of attack can be predicted in a short time.

3.1.2. Engine Model under the Influence of Inlet Distortion

The inlet distortion creates an uneven flow field at the engine inlet, which leads to stall boundary degradation and steady-state characteristic decay of the entire engine, causing the stall point of the entire compressor to move to the lower right [21] (Figure 4), that is to say, the inlet distortion affects the performance parameters of each component of the engine and results in the offset of the common working line and the reduction in the surge margin.

The inhomogeneous air flow caused by the inlet distortion changes the inlet total pressure and affects the thermodynamic process of the subsequent parts. The average total pressure recovery coefficient σ of the inlet cross-section is generally used to measure the inlet total pressure after the inlet distortion. The main formulas are as follows.

$$\sigma_{AV} = \frac{\int_0^{2\pi} \sigma_r(\theta) d\theta}{2\pi} \quad (38)$$

in the formula, $\sigma_r(\theta)$ is the radial mean total pressure recovery coefficient function, and the expression is

$$\sigma_r(\theta) = \frac{\int_{\bar{r}_{hub}}^{1} 2\sigma(\bar{r},\theta)\bar{r}d\theta}{1-\bar{r}_{hub}^2} \tag{39}$$

where $\bar{r}_{hub}$ is the relative radius of the wheel hub and $\sigma(\bar{r},\theta)$ is the circumferential angle of θ; the total pressure recovery coefficient is a function of the relative radius $\bar{r}$, and the schematic diagram of the parameters is shown in Figure 5. The shaded area is below σ_{AV} in the low-pressure region θ^-. θ_1 and θ_2 are the starting and ending points of the low-pressure region, respectively.

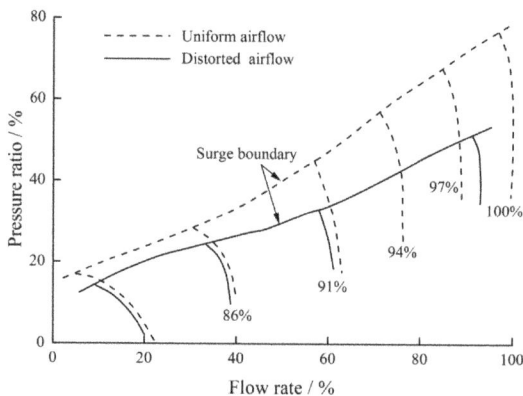

Figure 4. Sketch of the compressor surge boundary movement caused by inlet distortion.

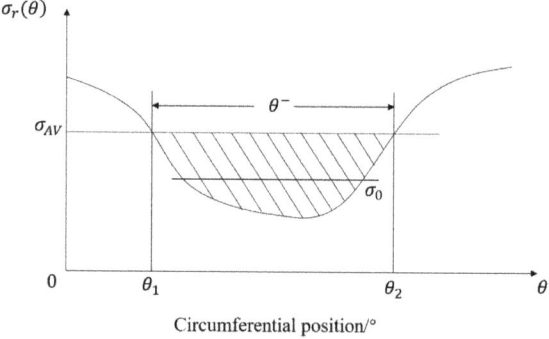

Figure 5. Radial mean total pressure recovery coefficient $\sigma_r(\theta)$.

The recovery coefficient of the surface mean total pressure directly reflects the variation in total pressure after distortion [17]. Figure 6 shows the surface mean total pressure recovery coefficient σ_{AV} and the mean total pressure σ_0 of the distorted low-pressure sector of an inlet at different angles of attack during subsonic and sonic travel. The angle of attack is directly related to the distortion. The larger the angle of attack, the stronger the distortion is. As can be seen from Figure 6, the total pressure recovery coefficient is closely related to the angle of attack. The σ_{AV} attenuates rapidly from the 0° angle of attack to both sides, and the attenuation is faster at a negative angle of attack. It can be concluded that σ_{AV} is closely related to the distortion strength and increases with the increase in distortion strength. Therefore, the inlet total pressure in the component-level model needs to be corrected according to the inlet distortion.

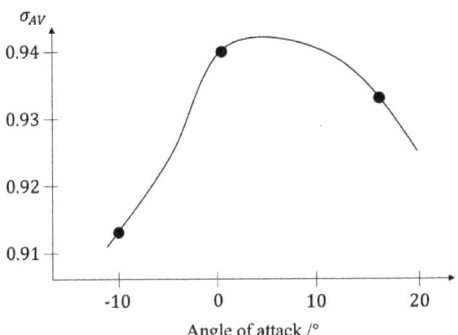

Figure 6. The surface mean total pressure recovery coefficient σ_{AV} of an inlet at different angles of attack during subsonic operation.

In order to quantify the impact of inlet distortion, it is necessary to select a distortion index that can reflect the magnitude of the distortion. The distortion index is mainly used to measure the degree of distortion quantitatively. At present, different countries use different calculation methods, so there is not a universal definition of the distortion index. Because the Russian comprehensive distortion index is widely used in engine stability assessment and experimental research, the research work in this paper is based on the definition of the distortion index. The distortion index only considers the total pressure distortion, not the total temperature distortion, and mainly consists of three parts, namely, the circumferential total pressure unevenness, the radial total pressure unevenness, and the mean surface turbulence. However, a large number of statistical data show that the influence of radial total pressure unevenness on the stability of different types of engine structures is very small, so it is not considered. Therefore, the total pressure distortion index can be expressed as

$$W = \Delta\overline{\sigma_0} + \varepsilon_{AV} \tag{40}$$

The composite distortion index W is the sum of the steady-state circumferential total pressure distortion index $\Delta\overline{\sigma_0}$ and the surface mean turbulence degree ε_{AV}. For the convenience of description, the Ws in this paper are all percentages. The $\Delta\overline{\sigma_0}$ indicates the difference between the surface mean total pressure recovery coefficient and the total pressure recovery coefficient in the low-pressure area.

$$\Delta\overline{\sigma_0} = 1 - \frac{\sigma_0}{\sigma_{AV}} \tag{41}$$

The mean surface turbulence ε_{AV} represents the quantitative characteristics of the total pressure fluctuation at the aerodynamic interface.

$$\varepsilon_{AV} = \frac{\sum_{i=1}^{n} \varepsilon_i}{n} \tag{42}$$

ε_i is the ratio of the root-mean-square value of the pulsating pressure to the mean value of the total pressure P^*.

$$\varepsilon_i = \Delta P^* / \overline{P^*}$$

$$\Delta P^* = \sqrt{\frac{1}{T} \int_0^T \left(P^*(\tau) - \overline{P^*} \right)^2} \tag{43}$$

$$\overline{P^*} = \frac{1}{T} \int_0^T P^*(\tau)\, d\tau$$

In the formula above, T is the sampling time of the pulse airflow.

In Section 2.2.1, we present the calculation of inlet and outlet flow parameters, and in Formula (4), we refer to the total inlet pressure recovery factor σ_i, which is used to characterize the total pressure loss of the inlet flow. The general total pressure recovery coefficient is calculated by the following formula.

$$\sigma_i = \begin{cases} 1.0 - 0.075(M_a - 1)^{1.35} & , M_a > 1.0 \\ 1.0 & , M_a \leq 1.0 \end{cases} \quad (44)$$

For the total pressure recovery coefficient with inlet distortion, it is correlated not only with M_a but also with W. The total pressure recovery coefficient in this paper is:

$$\sigma_i = \begin{cases} (1 - 0.01W)\left[1 - 0.075(M_a - 1)^{1.35}\right] & , M_a > 1.0 \\ 1 - 0.01W & , M_a \leq 1.0 \end{cases} \quad (45)$$

The conventional fan component calculation module obtains the conversion flow and efficiency based on the interpolation of the fan relative conversion speed, and the calculation expression is

$$\sigma_i = \begin{cases} (1 - 0.01W)\left[1 - 0.075(M_a - 1)^{1.35}\right] & , M_a > 1.0 \\ 1 - 0.01W & , M_a \leq 1.0 \end{cases} \quad (46)$$

Similarly, in Section 2.2.2, some modifications are made to the fan characteristics by taking into account the inlet distortion index W in the fan converted air flow and the overall efficiency calculations, i.e., Formula (6) is changed to:

$$\begin{aligned} m_{aL,cor} &= m_{aL,cor}(n_{cL,cor}, \pi_{cL}^*, \phi_L, W) \\ \eta_{cL}^* &= \eta_{cL}^*(n_{cL,cor}, \pi_{cL}^*, \phi_L, W) \end{aligned} \quad (47)$$

So far, the calculation flow of the components related to the inlet distortion has been modified and combined with the existing engine component-level modeling methods. The nonlinear engine component-level model with the inlet distortion can be formed, and the schematic diagram of its main calculation flow is shown in Figure 7.

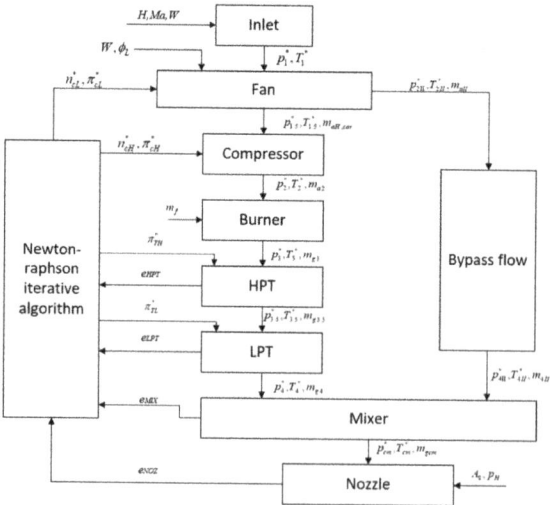

Figure 7. Flow chart of the turbine engine component-level model with inlet distortion.

3.2. Adaptive Disturbance Rejection Control

The basic configuration of adaptive disturbance rejection control is a nonlinear dynamic inverse (NDI) control law. The control logic is shown in Figure 8.

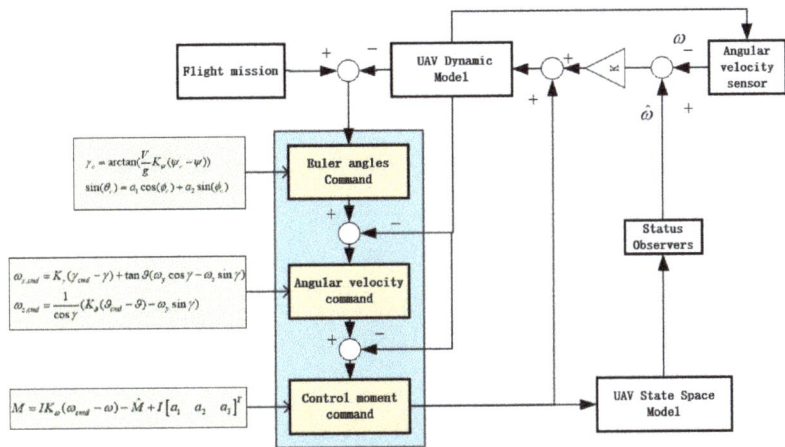

Figure 8. Control Structure chart of adaptive disturbance rejection.

Taking the tilt-turn dynamics model of UAV as an example, first, according to the dynamics of UAV established in Section 2.1, the instruction design rule is that the time scale of the state variable is separated, and the control instructions are generated by dynamic inversion according to the response speed of the variable.

3.2.1. Design of Slow Loop Control Instruction

The desired angular rate command $\dot{\gamma}_t$ and $\dot{\vartheta}_t$ can be obtained from the desired attitude angular response, and the expression is as follows:

$$\dot{\gamma}_t = K_\gamma (\gamma_{cmd} - \gamma)$$
$$\dot{\vartheta}_t = K_\vartheta (\vartheta_{cmd} - \vartheta) \tag{48}$$

The angular rate instructions can be obtained by associating with the dynamic equations. The results are as follows:

$$\omega_{x,cmd} = K_\gamma (\gamma_{cmd} - \gamma) + \tan \vartheta (\omega_y \cos \gamma - \omega_z \sin \gamma)$$
$$\omega_{z,cmd} = \frac{1}{\cos \gamma} \left(K_\vartheta (\vartheta_{cmd} - \vartheta) - \omega_y \sin \gamma \right) \tag{49}$$

Observing the dynamic model, the angular acceleration model can be considered as the result of the interaction of the moment part and the non-moment part, and the non-moment control parts are a_1, a_2, a_3.

$$a_1 = -\frac{(I_{zz} - I_{yy})I_{yy} - I_{xy}^2}{I_{xx}I_{yy} - I_{xy}^2} \omega_y \omega_z - \frac{(I_{xx} - I_{zz} + I_{yy})I_{xy}}{I_{xx}I_{yy} - I_{xy}^2} \omega_x \omega_z$$
$$a_2 = -\frac{(I_{xx} - I_{zz})I_{xx} - I_{xy}^2}{I_{xx}I_{yy} - I_{xy}^2} \omega_x \omega_z + \frac{(I_{xx} - I_{zz} + I_{yy})I_{xy}}{I_{xx}I_{yy} - I_{xy}^2} \omega_y \omega_z \tag{50}$$
$$a_1 = -\frac{I_{yy} - I_{xx}}{I_{zz}} \omega_x \omega_y + \frac{I_{xy}}{I_{zz}} \left(\omega_x^2 - \omega_y^2 \right)$$

The above equation can be rewritten as

$$\begin{bmatrix} \dot{\omega}_x \\ \dot{\omega}_y \\ \dot{\omega}_z \end{bmatrix} = \begin{bmatrix} -I_{xy} & I_{yy} & 0 \\ I_{xx} & -I_{xy} & 0 \\ 0 & 0 & I_{zz} \end{bmatrix}^{-1} \left(\begin{bmatrix} l \\ m \\ n \end{bmatrix} + \begin{bmatrix} l_\delta \\ m_\delta \\ n_\delta \end{bmatrix} \right) + \begin{bmatrix} a_1 \\ a_2 \\ a_3 \end{bmatrix} \quad (51)$$

The control quantities are $l_\delta, m_\delta, n_\delta$, and the control instruction is designed according to the above expected response method.

$$\begin{bmatrix} \dot{\omega}_{x,t} \\ \dot{\omega}_{y,t} \\ \dot{\omega}_{z,t} \end{bmatrix} = \begin{bmatrix} K_{\omega_x} & 0 & 0 \\ 0 & K_{\omega_y} & 0 \\ 0 & 0 & K_{\omega_z} \end{bmatrix} \begin{bmatrix} \omega_{x,cmd} - \omega_x \\ \omega_{y,cmd} - \omega_y \\ \omega_{z,cmd} - \omega_z \end{bmatrix} \quad (52)$$

which is associated with the kinetic equations:

$$\begin{bmatrix} l_{\delta,cmd} \\ m_{\delta,cmd} \\ n_{\delta,cmd} \end{bmatrix} = \begin{bmatrix} -I_{xy} & I_{yy} & 0 \\ I_{xx} & -I_{xy} & 0 \\ 0 & 0 & I_{zz} \end{bmatrix}^{-1} \left(\begin{bmatrix} K_{\omega_x} & 0 & 0 \\ 0 & K_{\omega_y} & 0 \\ 0 & 0 & K_{\omega_z} \end{bmatrix} \begin{bmatrix} \omega_{x,cmd} - \omega_x \\ \omega_{y,cmd} - \omega_y \\ \omega_{z,cmd} - \omega_z \end{bmatrix} - \begin{bmatrix} a_1 \\ a_2 \\ a_3 \end{bmatrix} \right) - \begin{bmatrix} l \\ m \\ n \end{bmatrix} \quad (53)$$

3.2.2. Design of Adaptive Disturbance Rejection Control

In the above derivation process, the body torque of the UAV is strongly nonlinear during the flight process, and it cannot accurately rely on the existing data to calculate it during the flight process. Similarly, when the moment control command is calculated, the rudder surface efficiency cannot be obtained accurately, and the exact value of rudder surface deflection cannot be determined to realize the control moment; the original control law is improved to overcome its dependence on an accurate model. The adaptive disturbance rejection control operates by comparing the measured response with an internal model of the desired dynamics. It then adjusts the control signal based on the difference between the desired and measured behaviors. First, the desired angular rate is designed as follows:

$$\dot{\omega}_{des} = K_\omega \omega_{cmd} - \omega_{des} \quad (54)$$

Because of the simplified assumptions, model uncertainty, external disturbance, and so on, dynamic inversion cannot achieve its goal completely. Therefore, d is introduced to describe all the uncertainty errors of the system. The system dynamics are written in a form similar to the expected behavior as follows.

$$\dot{\omega} = K_\omega(\omega_{cmd} - \omega_{des}) + u_{ad} + d \quad (55)$$

Among them, u_{ad} brings the angular acceleration increment for the adaptive control moment, $u_{ad} = I^{-1} \Delta M$. The angular rate internal model instruction is designed as.

$$\dot{\hat{\omega}} = K_\omega(\omega_{cmd} - \omega_{des}) + u_{ad} - K_d(\tilde{\omega} - \omega) \quad (56)$$

where $\hat{\omega}$ is an output of the UAV state space model. $\tilde{\omega} = \hat{\omega} - \omega$ is defined. The difference between the true angular rate and the internal model angular rate is obtained.

$$\dot{\tilde{\omega}} = \dot{\hat{\omega}} - \dot{\omega} = -K_{ad}\tilde{\omega} - d \quad (57)$$

According to the solution of the first-order non-homogeneous differential equation, the functional relationship of $\tilde{\omega}$ with time can be solved.

$$\tilde{\omega} = -\frac{d}{K_{ad}} + \left(\tilde{\omega}(0) + \frac{d}{K_{ad}} \right) e^{-K_{ad}t} \quad (58)$$

From the above, we know that, in a certain period of time, $\tilde{\omega}$ converges to $-d/K_{ad}$, and in the same way, in this period of time, $\dot{\tilde{\omega}}$ converges to 0, that is, $-K_{ad}\tilde{\omega} - d$ converges to 0. Based on the actual ω differential equation, the $u_{ad} = K_{ad}\tilde{\omega}$ can be substituted into Formula (61) to obtain:

$$\dot{\omega} = K_\omega(\omega_{cmd} - \omega_{des}) \tag{59}$$

According to the solution of the first-order non-homogeneous linear differential equation,

$$\omega = \omega_{cmd} + (\omega(0) - \omega_{cmd})e^{-K_{ad}t} \tag{60}$$

In this case, ω converges to ω_{cmd} within a certain time, so $u_{ad} = K_{ad}\tilde{\omega}$ can suppress the effect of error d.

According to the dynamic inversion of the above dynamic equations, the adaptive input is added to the NDI control torque command to generate the total torque command.

$$M = IK_\omega(\omega_{cmd} - \omega_{des}) + IK_{ad}(\hat{\omega} - \omega) + I\begin{bmatrix} a_1 & a_2 & a_3 \end{bmatrix}^T \tag{61}$$

4. Simulation Results

In the preceding paragraph, we build the dynamics model of UAV and the engine model of UAV and add the influence of inlet distortion into the engine model. Then, the strategy of thrust optimization based on the adaptive margin model and the method of adaptive disturbance rejection flight control for UAV are proposed. In order to verify this, the flight control strategy and the thrust optimization strategy proposed earlier are simulated and verified successively. The parameters of a certain type of UAV are given in Table 2. The simulation is based on the model and strategy established in the previous paper, which are implemented in code in MATLAB, where the turbine engine model is built by Simulink of MATLAB.

Table 2. Parameters of a certain type of UAV.

Parameter	Symbol	Value
aircraft mass (kg)	m	9295.44
wingspan (m)	b	9.144
wing area (m^2)	S	27.87
mean aerodynamic chord (m)	$\bar{c}$	3.45
roll moment of inertia (kg·m^2)	I_{xx}	12,874.8
yaw moment of inertia (kg·m^2)	I_{yy}	85,552.1
pitch moment of inertia (kg·m^2)	I_{zz}	75,673.6
product moment of inertia (kg·m^2)	I_{xy}	1331.4
product moment of inertia (kg·m^2)	I_{xz}	0
product moment of inertia (kg·m^2)	I_{yz}	0
c.g. location (m)	x_{cg}	$0.3\bar{c}$
reference c.g. location (m)	x_{cgr}	$0.35\bar{c}$

4.1. Simulation Results of Thrust Optimization Based on the Adaptive Margin Model

In order to verify the optimization results of the engine thrust performance, we use the acceleration performance improvement of the UAV at the altitude of 7500 m as a way to test the engine thrust optimization, to increase the speed of the UAV from 150 m/s to 250 m/s with the altitude change, and to check the time of the engine acceleration plan as well as the thrust output of the engine. The simulation results are as follows:

Figure 9a shows the speed simulation results of the UAV for a given speed command, and we compare the speed results after the thrust performance optimization with the original case. It can be seen that the acceleration performance of the UAV after the thrust lift is obviously improved compared with the original situation. In the original acceleration plan, the UAV used 28.9 s to accelerate the speed to 249 m/s, the speed was increased to 250 m/s in 40 s, the time of acceleration to 249 m/s was advanced to 25.5 s, and the time of acceleration to 250 m/s was reduced to 29.9 s after the thrust optimization. This shows that, under the same engine acceleration control parameters, the engine thrust optimization strategy proposed in this paper can significantly improve the acceleration capability of UAV and shorten the acceleration time. It is also possible to see this by the slope of the velocity resultant curve in plot 9(a), i.e., the maximum acceleration of the UAV.

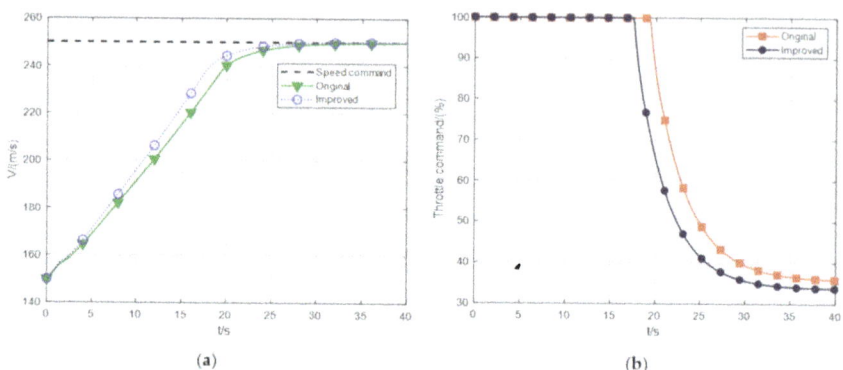

Figure 9. The result graph of engine thrust lifting: (**a**) Speed result of the UAV acceleration process, (**b**) Result of the UAV engine throttle command.

Figure 9b shows the result of UAV engine throttle command. Under the original condition, the engine throttle command needs to keep 19.3 s fully open, but after improvement, the change time is reduced to 17.5 s. The result shows that the full throttle time of UAV is shorter after the thrust performance is improved. In the acceleration control program of the UAV, the controller parameters used in the two simulated scenarios are exactly the same. Therefore, the optimized throttle command maximum output time is shorter than the original one, which can also indicate that the optimized case has an increased maximum thrust output compared to the original case.

To visualize the thrust improvement, Figure 10 shows the UAV thrust output during this acceleration; the percentage increase in the thrust of the engine at 100% throttle is in the corner of the image (take the first 17.5 s, which is 100% of the time for the engine throttle command in both cases). The right axis of the small graph is the label, and the unit is the percentage of thrust performance improvement. It can be seen that the maximum thrust output of the engine can be increased by at least 8% under the condition of maximum throttle command and also increases with the increase in the UAV speed, reaching 10.4% at 17.5 s.

Figure 11 is the result diagram of the working state of the compressor. It can be seen that the working point of the engine is closer to the surge boundary after the more reasonable order of the turbocharging ratio, that is, the surge margin reserve of the engine is reduced but still kept in safe working conditions, and the engine did not surge.

Brief summary:

1. Under the same acceleration control parameters, the thrust optimization strategy proposed in this paper can significantly improve the acceleration capability of UAV and shorten the acceleration time.
2. With improved thrust performance, the UAV requires a shorter full throttle time.
3. At the same flat flying speed, the throttle command of the engine after thrust optimization is smaller than the original throttle command.
4. At maximum throttle, the engine's maximum thrust output can be increased by at least 8%.

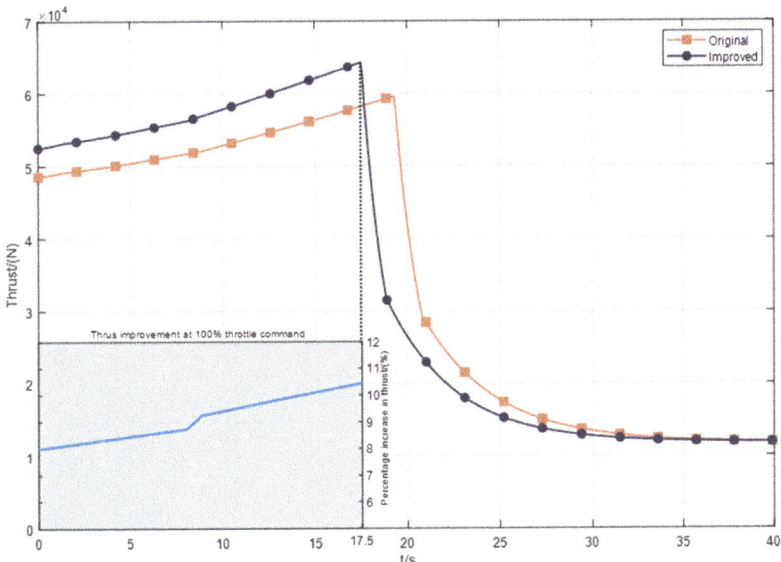

Figure 10. Thrust output diagram.

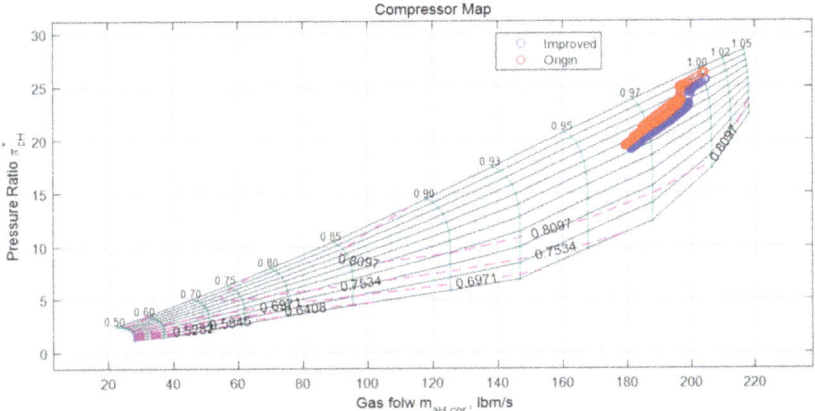

Figure 11. Compressor working state diagram.

4.2. Simulation Results of Adaptive Disturbance Rejection Control

In order to verify the designed adaptive disturbance rejection control strategy, the UAV is given different flight instructions in the longitudinal and rolling directions, and the advantages and disadvantages of the control method and the traditional PID control were observed. The flight environment is given under two conditions: altitude 7500 m, speed 150 m/s.

The first is the longitudinal simulation results, the initial pitch angle is $\gamma = 7.2°$, the initial track angle α is $0°$, the UAV is controlled in the first 5 s of the flat flight, the UAV is given a $15°$ pitch command at 5 s, and the UAV is made to climb quickly. Observe the response of the pitch angle and the command output of the rudder deviation.

For the PID control case in the simulation, we used the Ziegler–Nichols tuning method. This method provides a good starting point for the PID gains, which can then be refined through further tuning. Start by adjusting the proportional gain Kp until the system's response is close to the desired response. Then, adjust the integral gain Ki until the steady-state error is reduced to an acceptable level. Finally, adjust the derivative gain Kd to reduce the overshoot and dampen the oscillations caused by the proportional and integral gains. A set of PID controller parameters with the fastest response time and the smallest overshoot is obtained by the above method.

As shown in Figure 12c, it took 17.39 s for the PID controller to make the pitch angle converge to the command, while the adaptive disturbance rejection control took 3.62 s to achieve the equivalent situation (0.05 degrees of static difference). Compared with the PID control, the pitch angle of the adaptive disturbance rejection control method converges to the given instruction more rapidly, the overshoot is smaller, and the overshoot arrival time is shorter.

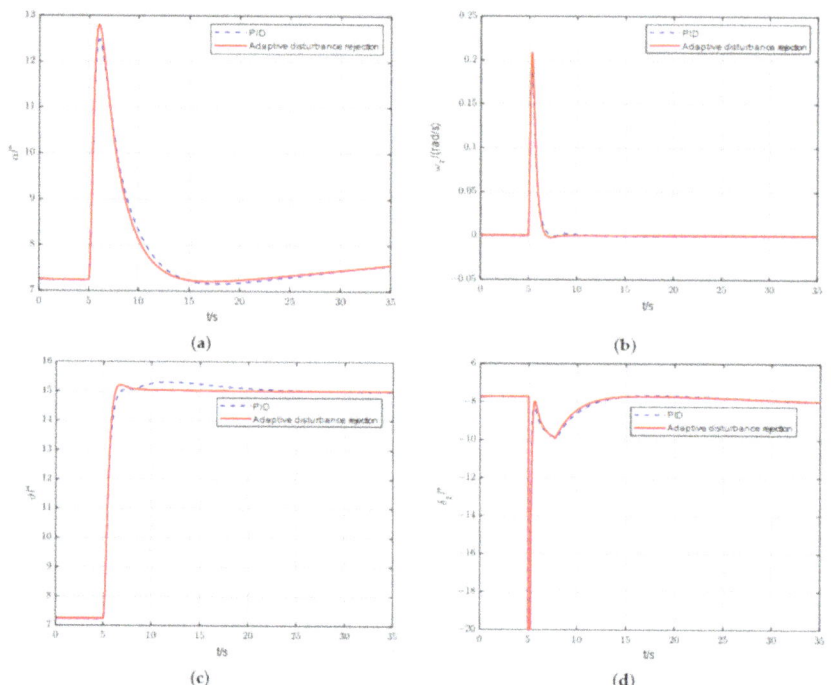

Figure 12. Simulation results of UAV adaptive disturbance rejection control in longitudinal flight: (**a**) Result of the angle of attack response graph; (**b**) Result of the pitch angular velocity response graph; (**c**) Result of the pitch angular response graph; (**d**) Result graph of the elevator deflection command.

Similarly, the simulation results for the roll direction are given in Figure 13. The initial tilt angle is 0°, and the control UAV will fly horizontally in the first 5 s. The UAV is given a 35° tilt command at 5 s to make the UAV roll quickly. Observe the response of the tilt angle and the command output of the rudder.

As shown in Figure 13a, The PID controller produces a large overshoot in the tilt angle (12.82% overshoot), which is obviously undesirable in this case, while the adaptive disturbance rejection control method has only 0.4% overshoot. Compared to the result of PID control, the tilt angle of the adaptive disturbance rejection control mode converges to the given instruction more quickly, the overshoot is smaller, and the overshoot arrival time is shorter. The aileron command, compared with the PID control in the case of swing, greatly reduced. This increases the potential of the UAV to perform other maneuvers. The results of both the longitudinal and rolling flight simulation show that the adaptive disturbance rejection control method proposed in this paper can achieve fast attitude maneuvers; compared with PID control, it can complete the maneuver command more accurately in a short time and improve the overshoot and rudder surface jitter caused by PID control.

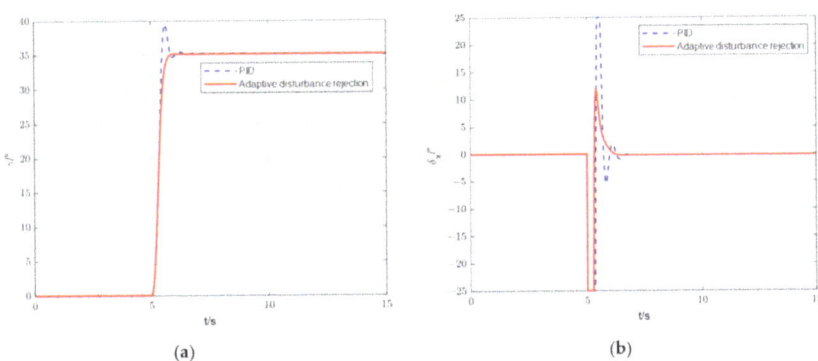

Figure 13. Flight simulation results of UAV adaptive interference suppression control roll direction: (**a**) The resulting graph of the tilt-angle response; (**b**) Result diagram of aileron deflection command.

5. Conclusions

We designed a strategy for exchanging the surge margin of the engine for the improvement of the thrust performance in order to meet the requirement of the development of the propulsion system for UAVs. First, the dynamic equation of UAV is established, which describes the straight motion of the center of mass and the rotation motion of the body around the center of mass. A mathematical model of the turbofan's components is built. Then, an optimal thrust algorithm based on the adaptive margin model is designed. The internal working environment of the engine and the angle of attack in flight dynamics are considered. From the angle of aircraft/engine integration, the way to improve the performance of the engine propulsion system is studied; in this part, the UAV body transfer function is obtained from the UAV dynamic equation, and the UAV angle of attack is predicted by the transfer function. Then, the inlet and fan models are modified by adding a correction term considering inlet distortion. For further flight simulation, a flight control strategy based on adaptive disturbance rejection control is designed. In order to compensate for the external loop control law, a disturbance observer is introduced to estimate the flight disturbance of the UAV.

Finally, the feasibility of the proposed strategy is verified by numerical simulation. The numerical simulation results show that the thrust lifting strategy based on the adaptive margin model can ensure the safety of the flight and the stability of the compressor; the maximum thrust of the UAV is increased by at least 8% by reducing the surge margin reserve and improving the maneuverability of the UAV. The adaptive disturbance rejection control method can suppress the attitude fluctuation quickly and has strong robustness and adaptive ability.

Nowadays, there are more and more research works on aircraft/engine integration, and in this paper, we meet the demand of the thrust performance improvement of UAVs from this perspective, which has some implementability in engineering practice. However, the strategy still has some limitations, namely, a high accuracy requirement for the engine model. In fact, turbine engine power system modeling is a complex discipline, and the engine model has strong nonlinearity. The component-level modeling method used in this paper, at the cost of a lengthy and complex mechanism model and in exchange for a higher accuracy, sacrifices a certain amount of computing speed. We plan to continue to explore the thrust optimization method based on the adaptive margin model and, on the other hand, to seek a more accurate and faster engine modeling method.

Author Contributions: Conceptualization, Y.W. and H.L.; methodology, Y.W.; software, K.L.; validation, K.L. and H.L.; formal analysis, Y.W., H.L. and K.L.; investigation, H.L.; resources, Y.W.; data curation, Y.W. and H.L.; writing—original draft preparation, H.L. and K.L.; writing—review and editing, Y.W. and H.L.; visualization, Y.W. and H.L.; supervision, Y.W.; project administration, Y.W.; funding acquisition, K.L. All authors have read and agreed to the published version of the manuscript.

Funding: This research was funded by the National Natural Science Foundation of China, grant number U2141229, and the foundation under grant JCJQ, grant number 2019-JCJQ-DA-001-131.

Data Availability Statement: Not applicable.

Conflicts of Interest: The authors declare no conflict of interest.

References

1. Bello, A.B.; Navarro, F.; Raposo, J.; Miranda, M.; Zazo, A.; Álvarez, M. Fixed-Wing UAV Flight Operation under Harsh Weather Conditions: A Case Study in Livingston Island Glaciers, Antarctica. *Drones* **2022**, *6*, 384. [CrossRef]
2. Cao, S.; Yu, H. An Adaptive Control Framework for the Autonomous Aerobatic Maneuvers of Fixed-Wing Unmanned Aerial Vehicle. *Drones* **2022**, *6*, 316. [CrossRef]
3. Liu, Z.; Han, W.; Wu, Y.; Su, X.; Guo, F. Automated Sortie Scheduling Optimization for Fixed-Wing Unmanned Carrier Aircraft and Unmanned Carrier Helicopter Mixed Fleet Based on Offshore Platform. *Drones* **2022**, *6*, 375. [CrossRef]
4. Ming, Z. Research on Uav Application Development and Airworthiness Standard Management in China. *Acad. J. Eng. Technol. Sci.* **2020**, *4*, 37–46.
5. Cui, Y.P.; Xing, Q.H. The challenge and inspiration of UAVs to field air defense from the Russia-Ukraine War. *Aerosp. Electron. Warf.* **2022**, *38*, 1–3.
6. Li, R.J.; Gong, S. Development Direction Analysis of Future UCAV Power Plant. *Aeronaut. Sci. Technol.* **2015**, *26*, 1–5.
7. Hu, X.Y. UAV Propulsion System Technology Research. *Gas Turbine Exp. Res.* **2008**, *1*, 58–61.
8. Hu, X.Y. Development of High Altitude Long Endurance UAV Propulsion Technology. *Gas Turbine Exp. Res.* **2006**, *4*, 56–60.
9. Chen, D.; Tian, H.J.; Li, M.X.; Lei, Q.Q. Analysis to the Development of Power Systems for MALE UAVs. *Aerosp. Power* **2022**, *5*, 16–19.
10. Cheng, B.L.; Tang, H.; Xu, X.; Li, X.P. Performance Study on Turbofan Engine with Turbine Inter Burner. *Aeroengine* **2010**, *36*, 18–22.
11. Dong, H.B.; Shang, G.J.; Guo, Y.Q. Effect and verification of mass injecting pre-compressor cooling on control plan of turbonfan engine. *J. Aerosp. Power* **2022**, *37*, 404–408.
12. Sun, H.G.; Li, J.; Li, H.C.; Ye, B. The Analysis on Surge Margin'S Influence on Aeroengine Thrust Control System. *Comput. Simul.* **2017**, *34*, 84–89.
13. Longley, J.P. A review of non-steady flow models for compressor stability. *J. Turbomach.* **1994**, *116*, 202–215. [CrossRef]
14. Longley, J.P.; Shin, H.W.; Plumley, R.E. Effects of rotating inlet distortion on multistage compressor stability. *ASME J. Turbomach.* **1996**, *118*, 181–188. [CrossRef]
15. Wang, J.K.; Zhang, H.B.; Chen, K.; Sun, F.Y.; Zhou, X. High stability control of engine based on surge margin estimation model. *J. Aerosp. Power* **2013**, *28*, 2145–2154.
16. Liu, D.X. *Aviation Gas turbine Engine Stability Design and Evaluation Technology*, 6th ed.; Aviation Industry Press: Beijing, China, 2004; pp. 45–182.
17. Bodó, Z.; Lantos, B. Modeling and control of fixed wing UAVs. In Proceedings of the IEEE 13th International Symposium on Applied Computational Intelligence and Informatics (SACI), Timisoara, Romania, 29–31 May 2019; pp. 332–337.
18. Kim, C. Development of unified high-fidelity flight dynamic modeling technique for unmanned compound aircraft. *Int. J. Aerosp. Eng.* **2021**, *2021*, 5513337.
19. Zhou, W.X. *Research on Object-Oriented Modeling and Simulation for Aeroengine and Control System*; DR, Nanjing University of Aeronautics and Astronautics the Graduate School: Nanjing, China, 2006.
20. Iliff, K.W. *Flight-Determined Subsonic Longitudinal Stability and Control Derivatives of the F-18 High Angle of Attack Research Vehicle (HARV) with Thrust Vectoring*; NASA, Dryden Flight Research Center: Edwards, CA, USA, 1997; pp. 17–19.
21. Lee, K.; Lee, B.; Kang, S.; Yang, S.; Lee, D. Inlet Distortion Test with Gas Turbine Engine in the Altitude Engine Test Facility. In Proceedings of the 27th AIAA Aerodynamic Measurement Technology and Ground Testing Conference, Chicago, IL, USA, 30 June 2010; p. 4337.

Disclaimer/Publisher's Note: The statements, opinions and data contained in all publications are solely those of the individual author(s) and contributor(s) and not of MDPI and/or the editor(s). MDPI and/or the editor(s) disclaim responsibility for any injury to people or property resulting from any ideas, methods, instructions or products referred to in the content.

Article

Finite-Time Contractive Control of Spacecraft Rendezvous System

Jing Sheng [1], Yunhai Geng [1,*], Min Li [2] and Baolong Zhu [3]

[1] School of Astronautics, Harbin Institute of Technology, Harbin 150001, China
[2] Beijing Institute of Control Engineering, Beijing 100190, China; liminhit@126.com
[3] School of Information and Automation, Qilu University of Technology, Jinan 250353, China
* Correspondence: gengyh@hit.edu.cn

Abstract: In this paper we investigate the problem of a finite-time contractive control method for a spacecraft rendezvous control system. The dynamic model of relative motion is formulated by the C-W equations. To improve the convergent performance of the spacecraft rendezvous control system, a finite-time contractive control law is introduced. Lyapunov's direct method is employed to obtain the existence condition of the desired controllers. The controller parameter can be obtained with the help of the cone complementary linearization algorithm. A numerical example verifies the effectiveness of the obtained theoretical results.

Keywords: finite-time contractive stability; state feedback control; spacecraft rendezvous system; cone complementary linearization; C-W equations

MSC: 93D40; 70M99

1. Introduction

The spacecraft rendezvous system is an important part of the orbital spacecraft since it provides important technical support for various space missions such as astronaut pick-up, material supply, space station construction and maintenance, and even manned lunar landings and deep-space exploration missions. An autonomous rendezvous system involves two spacecraft: one is the target spacecraft and the other is the chaser spacecraft. In general, the relative dynamic model of two spacecraft is a set of nonlinear equations [1]. To facilitate analysis and controller design, two kinds of linearized relative motion models were developed, namely the Clohessy–Wiltshire (C-W) equation [2] and the Tschauner–Hempel (T-H) equation [3]. The C-W equation is linear time-invariant and is suitable for target spacecraft running in circular orbits. In contrast, the T-H equation is linear time-varying and is more appropriate for target spacecraft operating in circular orbits.

The quality of the adopted control strategies directly affects the overall performance of the autonomous rendezvous system, and then affects the orbital service mission. This has stimulated an outpouring of enthusiasm from researchers and in the past decades, various insightful and innovative results on the control of the autonomous rendezvous of spacecraft have emerged [4–10]. Here, to name a few, a new relative dynamic model that takes the parameter uncertainty and output tracking into account was developed in [5], and the guaranteed cost output tracking controller was designed by virtue of the convex optimization method and the linear matrix inequality technique. Moreover, saturated state feedback controllers were developed by Luo [7] to globally stabilize the spacecraft rendezvous system constrained by thrust saturation and/or time delay. In addition, the semi-global finite-time stabilization issue of a spacecraft rendezvous system with input constraints was reported in [8], where the dynamic event-triggered control and self-triggered control techniques were considered.

However, finite-time contractive stability (FTCS), proposed in [11] for the first time relates to the transient performance of systems in a fixed time interval rather than the steady performance over an infinite time interval. Roughly speaking, if, given the bound of the initial condition c_1, the state trajectory of a finite-time contractively stable system does not exceed a bound $c_2 > c_1$ over the prescribed time interval $[0, T_u]$, the state trajectory will further lie within a bound c_3 over the time interval $[t_s, T_u]$, and it will never escape from the bound c_3 after it comes in [12], where $0 < c_3 < c_1 < c_2$, and $0 < t_s < T_u$. This suggests that systems under FTCS also have superior convergence performance on the basis of "boundedness" [13]. In recent years, FTCS has drawn more attention, which has resulted in the FTCS issue of several kinds of systems being discussed, such as the stochastic system [14], impulsive systems [15,16], switched systems [17,18], and so on. Physical applications of FTCS in fields such as clinical medicine [19] and population control [20] have also been reported. Moreover, we note that there exists the potential practical application of finite-time contractive stability control of the spacecraft rendezvous systems on occasions where the relative distance and relative velocity along the x-axis, y-axis, and z-axis between the target spacecraft and chaser spacecraft need to be within an ideal prescribed bound after a fixed time t_s. However, to the best of the authors' knowledge, there exist few results on the FTCS of spacecraft rendezvous systems in the literature, which motivates this work.

The finite-time contractive control issue for a spacecraft rendezvous system is considered in this paper. The state feedback controller is designed to finite-time contractively stabilize the spacecraft rendezvous system. The main contribution of this paper is threefold as follows. (1) This is the first attempt for the finite-time contractive control of a spacecraft rendezvous system, and a sufficient condition for the existence of desired controllers is established. (2) A convex optimization problem with linear matrix inequality constraints is established for control synthesis, which can be solved by a cone complementary linearization algorithm. (3) A numerical example shows that the proposed controller has faster convergence speed compared with traditional control methods.

Notations: $tr(A)$ represents the trace of A. Matrix $A > 0 \ (\geq 0)$ denotes that A is a positive definite matrix (positive semi-definite matrix). Moreover, we assume that the dimensions of the matrices are compatible with each other, if this is not explicitly stated before. "w.r.t" denotes the phrase "with respect to".

2. Problem Formulation

We assume that the target spacecraft is running in a circular orbit, and the coordinate frame for the two spacecraft is shown in Figure 1. The origin of the coordinate system is located at the center of mass of the target spacecraft. The x-axis is in the orbital plane of the target spacecraft, with the positive direction of the Earth center pointing to the target spacecraft. The y-axis points to the running direction of the target spacecraft. The z-axis is perpendicular to the orbital plane and forms a right-handed rectangular coordinate system with the other two axes. Hence, the relative dynamic motion would obey the following C-W equations [21]

$$\begin{cases} \ddot{x} - 2n\dot{y} - 3n^2 x = \frac{1}{m}T_x, \\ \ddot{y} + 2n\dot{x} = \frac{1}{m}T_y, \\ \ddot{z} + n^2 z = \frac{1}{m}T_z, \end{cases} \quad (1)$$

where x, y, and z stand for the relative position, m is the mass of the chaser, n is the angular velocity of the target spacecraft, and $T_i (i = x, y, z)$ is the i-th component of the specific control force acting on the relative motion dynamics. Letting $x(t) = [x, y, z, \dot{x}, \dot{y}, \dot{z}]^T$, and $u(t) = [T_x, T_y, T_z]^T$, then (1) can be further described as

$$\begin{cases} \dot{x}(t) = Ax(t) + Bu(t), \\ y(t) = Cx(t), \end{cases} \quad (2)$$

where $A = \begin{bmatrix} 0 & 0 & 0 & 1 & 0 & 0 \\ 0 & 0 & 0 & 0 & 1 & 0 \\ 0 & 0 & 0 & 0 & 0 & 1 \\ 3n^2 & 0 & 0 & 0 & 2n & 0 \\ 0 & 0 & 0 & -2n & 0 & 0 \\ 0 & 0 & -n^2 & 0 & 0 & 0 \end{bmatrix}$, $B = \begin{bmatrix} 0 & 0 & 0 \\ 0 & 0 & 0 \\ 0 & 0 & 0 \\ \frac{1}{m} & 0 & 0 \\ 0 & \frac{1}{m} & 0 \\ 0 & 0 & \frac{1}{m} \end{bmatrix}$, $C = \begin{bmatrix} 1 & 0 & 0 & 0 & 0 & 0 \\ 0 & 1 & 0 & 0 & 0 & 0 \\ 0 & 0 & 1 & 0 & 0 & 0 \end{bmatrix}$.

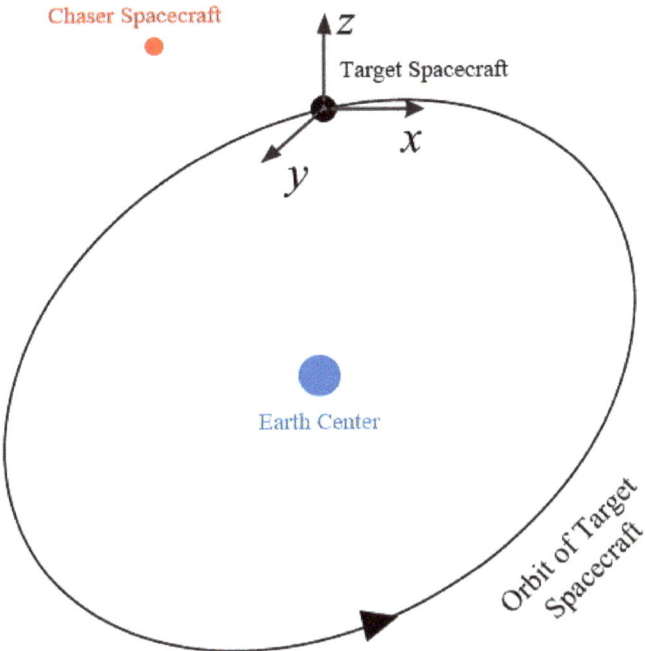

Figure 1. Coordinate frame.

Lemma 1 ([22]). *For matrices $P > 0$ and $H > 0$, if and only if the conditions*

$$tr(PH) = n, \tag{3}$$

$$\begin{bmatrix} P & I \\ I & H \end{bmatrix} \geq 0, \tag{4}$$

hold, $PH = I$ holds.

Definition 1 ([19]). *System (2) is finite-time contractively stable w.r.t $(c_1, c_2, c_3, R, t_s, T_u)$, if $x^T(0)Rx(0) < c_1$ implies that $x^T(t)Rx(t) < c_2$, $\forall t \in [0, T_u]$; furthermore, $x^T(t)Rx(t) < c_3$, $\forall t \in [t_s, T_u]$, where $0 < c_3 < c_1 < c_2$, $0 < t_s < T_u$, and $R > 0$.*

3. Finite-Time Contractive Stabilization

Consider a state feedback control law for (2)

$$u = Kx(t), \tag{5}$$

where K is the controller parameter to be designed. Then, the closed-loop system is established as below

$$\dot{x}(t) = (A + BK)x(t). \tag{6}$$

The following theorem gives a sufficient condition for the existence of state feedback controller (5) under which the closed-loop system (6) is finite-time contractively stable.

Theorem 1. *For given scalars $\alpha > 0$, $c_3 < c_1 < c_2$, and $0 < t_s < T_u$, and a matrix $R > 0$, the closed-loop system (6) is finite-time contractively stable w.r.t $(c_1, c_2, c_3, R, t_s, T_u)$, if there exist a symmetric matrix $P > 0$ and a matrix K, as well as a scalar $\varepsilon > 0$ satisfying*

$$PA + PBK + A^T P + K^T B^T P + \alpha P < 0, \tag{7}$$

$$R < P < \varepsilon R, \tag{8}$$

$$\varepsilon c_1 < c_2, \tag{9}$$

$$e^{-\alpha t_s} \varepsilon c_1 < c_3, \tag{10}$$

and the equation restriction

$$PH = I, \tag{11}$$

Proof. Choosing a Lyapunov function $V(x(t)) = x^T(t) P x(t)$, then taking the time derivative yields

$$\dot{V}(x(t)) = x^T(t)((PA + PBK) + (PA + PBK)^T)x(t). \tag{12}$$

Furthermore, according to (7), it can be obtained that

$$PA + PBK + A^T P + K^T B^T P < -\alpha P, \tag{13}$$

from which we have

$$x^T(t)((PA + PBK) + (PA + PBK)^T)x(t) < -\alpha x^T(t) P x(t), \tag{14}$$

i.e.,

$$\dot{V}(x(t)) < -\alpha V(x(t)). \tag{15}$$

Multiplying both sides of (15) by $e^{\alpha t}$, and then integrating both sides of it from 0 to t for $t \in [0, T_u]$, one has

$$V(x(t)) < e^{-\alpha t} V(x(0)). \tag{16}$$

Furthermore, since it yields from (8) that $x^T(t) R x(t) < V(x(t)) < \varepsilon x^T(t) R x(t)$, then, by letting $x^T(0) R x(0) < c_1$, it can be obtained from (16) and (9) that

$$x^T(t) R x(t) < V(x(t)) < \varepsilon c_1 < c_2, \quad \forall t \in [0, T_u]. \tag{17}$$

Simliar to the proof processes (16)–(17), by (8) and (10), it follows from (15) that

$$x^T(t) R x(t) < e^{-\alpha t_s} \varepsilon c_1 < c_3, \quad \forall t \in [t_s, T_u]. \tag{18}$$

Hence, according to Definition 1, system (6) is finite-time contractively stable w.r.t $(c_1, c_2, c_3, R, t_s, T_u)$. This completes the proof. □

Remark 1. *The parameters c_1, c_2, and c_3, where $0 < c_3 < c_1 < c_2$, represent the specific bounds within which system state variables lie over the prescribed time interval. They are generally chosen from practical consideration and are pre-specified in a given problem, as stated in [23,24]. Furthermore, the obtained conditions for the finite-time contractive stability control issue in theorems are provided in terms of feasibility problems [25]. Hence, this suggests that the expected parameters c_1, c_2, and c_3 that we choose are achievable and the considered system can be said to be finite-time contractively stable w.r.t. $(c_1, c_2, c_3, R, t_s, T_u)$ over the fixed time interval according to Definition 1* if the established sufficient conditions in theorems are feasible. In addition, if needed, achievable values of c_1, c_2, and c_3 that make the obtained sufficient conditions feasible can be chosen by using the one-dimension linear search method or trial-and-error method.

A sufficient condition for the existence of the finite-time contractive controller (5) is established in Theorem 1. However, it is different to solve the controller parameter K straightforwardly since there exists the nonlinear term PBK in inequality (7). To make the controller design numerically tractable, a controller design method is developed by the following theorem where the parameters P and K are separated.

Theorem 2. *For given scalars $\alpha > 0$, $c_3 < c_1 < c_2$, and $0 < t_s < T_u$, and a matrix $R > 0$, the closed-loop system (6) is said to be finite-time contractively stable w.r.t $(c_1, c_2, c_3, R, t_s, T_u)$, if there exist a matrix Q, symmetric matrices $H > 0$, $P > 0$, and a scalar $\varepsilon > 0$ such that*

$$(AH + BQ) + (AH + BQ)^T + \alpha H < 0, \tag{19}$$

$$R < P < \varepsilon R, \tag{20}$$

$$\varepsilon c_1 < c_2, \tag{21}$$

$$e^{-\alpha t_s} \varepsilon c_1 < c_3, \tag{22}$$

with the equation restriction

$$PH = I, \tag{23}$$

where $H = P^{-1}$ and the controller parameter is obtained by $K = QH^{-1}$.

Proof. Pre-and post-multiplying (19) by P, one has

$$(PA + PBK) + (PA + PBK)^T + \alpha P < 0. \tag{24}$$

Then, following from the proof processes of Theorem 1, it can be easily obtained that system (6) is finite-time contractively stable w.r.t $(c_1, c_2, c_3, R, t_s, T_u)$. Here, the proof is omitted for simplicity. □

Remark 2. *It follows from (19) that $\dot{V}(x(t)) < -\alpha V(x(t)) < 0$, which indicates that the system (6) must be Lyapunov asymptotically (exponentially) stable in the case of finite-time contractive stability control. Furthermore, due to the existence of contraction conditions (21) and (22) over a finite-time interval for state trajectory under finite-time contractive stability control, when the system (6) is Lyapunov asymptotically (exponentially) stable, it may not be finite-time contractively stable w.r.t prescribed parameters c_1, c_2, c_3, R, t_s, and T_u. Briefly speaking, if a system is said to be finite-time contractive stable, it must be Lyapunov asymptotically stable, while, conversely, it may not be. In addition, with the aim of small t_s and c_3, the convergence speed of finite-time contractively stable systems may be better than that of Lyapunov asymptotic stable systems, which results in the considered system approaching the equilibrium state faster under FTCS.*

Remark 3. *In Theorem 2, the analytic solution of controller parameters K is given in the form of $K = QH^{-1}$, which is numerically solvable through the use of the well-established variable substitution method. However, matrices P and H that only satisfy the conditions (19)–(22) may not qualify since the potential relationship shown in (23) does not hold in this case. Hence, to ensure that the obtained feasible set satisfies both the constraints (19)–(22) and (23), the following minimization problem is considered.*

Problem 1.

$$\min \; tr(PH)$$
$$s.t. \; (4) \; and \; (19)\text{–}(22)$$

Remark 4. *On one hand, according to Lemma 1, $tr(PH) \geq n$ always holds if (4) holds. Then, if and only if $tr(PH) = n$, $tr(PH)$ reaches the minimum, and $PH = I$ holds. Hence, conditions (19)–(23) are feasible, and the controller parameter K can be further solved, when the solution of Problem 1 is n. On the other hand, Problem 1 is essentially a non-convex problem and it is difficult to solve. Hence, inspired by [26], the following cone complementary linearization algorithm (CCLA) is employed to address it (Algorithm 1). By this algorithm, $\phi + tr(P_1 H + P H_1)$ is used to linearly approximate $tr(PH)$ at a given point (P_1, H_1), where ϕ is a constant that is small enough. In this way, if and only if $tr(P_1 H + P H_1) = 2n$, the constraint $PH = I$ holds.*

Algorithm 1 CCLA for solving Problem 1

Step 1. Given parameters α, c_1, c_2, c_3, R, t_s, and T_u. Moreover, set $j = 1$, $\phi = 1 \times 10^{-6}$, and maximum iterations $Iter = 50$.
Step 2. Compute conditions (4) and (19)–(22). If not feasible, exit; otherwise, go to **Step 3**.
Step 3. Set $(H_j, P_j, Q_j, \varepsilon_i) = (H, P, Q, \varepsilon)$, where (H, P, Q, ε) is the feasible solution attained in **Step 2**. Furthermore, compute Problem 1.
Step 4. Compare the value of $tr(P_i H + P H_i)$ with $2n$, where n is the dimension of P. If $|tr(P_i H + P H_i) - 2n| < \phi$, output the value of $K = Q H^{-1}$ and then exit; else, $j = j + 1$, and compare j with $Itea$, if $i \leq Iter$, go to **Step 3**; else, exit.

Next, the Lyapunov asymptotical stabilization and the classical linear quadratic regulator (LQR) control issues of system (2) are also discussed for comparison.

(A) Lyapunov asymptotical stabilization

When $\alpha = 0$, one has from (19) that

$$(AH + BQ) + (AH + BQ)^T < 0, \quad (25)$$

Then, based on Problem 1 and Remark 2, it can be attained that closed-loop system (6) is Lyapunov asymptotic stable (LAS) if the following Problem 2 is feasible.

Problem 2.

$$\min \; tr(PH)$$
$$s.t. \; (4) \; and \; (25)$$

The following Algorithm 2 can be applied to compute the above Problem 2.

Algorithm 2 CCLA for solving Problem 2

Step 1. Set $j = 1$, $\phi = 1 \times 10^{-6}$, and maximum iterations $Iter = 50$.
Step 2. Solve the conditions (4) and (25). If not feasible, exit; otherwise, go to **Step 3**.
Step 3. Set $(H_j, P_j, Q_j) = (H, P, Q)$, where (H, P, Q) is the feasible solution attained in **Step 2**. Furthermore, solve Problem 2.
Step 4. Compare the value of $tr(P_i H + P H_i)$ with $2n$, where n is the dimension of P. If $|tr(P_i H + P H_i) - 2n| < \phi$, output the value of $K = Q H^{-1}$ and then exit; else, $j = j + 1$, and compare j with $Itea$, if $i \leq Iter$, go to **Step 3**; else, exit.

(B) LQR control [27]

Considering system (2) with controllable (A, B), we can obtain an optimal LQR by using the full state feedback control law $u = -Kx$, which can minimize the performance index as below

$$J = \int_0^\infty \left(x^T \mathcal{Q} x + u^T \mathcal{R} u \right) dt \tag{26}$$

where symmetrical matrices $\mathcal{Q} \geq 0$ and $\mathcal{R} > 0$.

In this case, the controller gain K is represented as $K = \mathcal{R}^{-1} B^T P$, where P is the solution of the following algebraic Riccati equation

$$PA + A^T P + \mathcal{Q} - PB\mathcal{R}^{-1}B^T P = 0$$

4. Simulation Results

In this section, the effectiveness of the proposed method is verified through the use of the following example.

We assume that the mass m of the chaser spacecraft is 300 kg and the angular velocity n of the target spacecraft is 1.168×10^{-3} rad/s. Furthermore, we assume that the two spacecraft are relatively static at $t = 0$, and the initial relative positions of the two spacecraft are 750 m (along the x-axis), 650 m (along the y-axis), and 550 m (along the z-axis) at $t = 0$. Then, it is obtained that $x(0) = [750, 650, 550, 0, 0, 0]^T$. Next, we will stabilize the considered spacecraft rendezvous system in the case of finite-time contractive stability and the case of the Lyapunov asymptotical stability, respectively.

Case 1. Finite-time contractive stabilization

For given parameters $c_1 = 1.3 \times 10^6$, $c_2 = 2.5 \times 10^6$, $c_3 = 1 \times 10^4$, $R = I$, $t_s = 10$, and $T_u = 40$, we solve Algorithm 1 through the use of the Yalmip toolbox [28]; when $\alpha = 0.56$, it can obtain the following feasible set of Problem 1 as follows.

$$P = \begin{bmatrix} 1.4615 & 0.0046 & 0.0031 & 0.3576 & -0.0014 & -0.0010 \\ 0.0046 & 1.4644 & -0.0011 & -0.0013 & 0.3568 & 0.0004 \\ 0.0031 & -0.0011 & 1.4663 & -0.0008 & 0.0003 & 0.3562 \\ 0.3576 & -0.0013 & -0.0008 & 1.2772 & -0.0048 & -0.0033 \\ -0.0014 & 0.3568 & 0.0003 & -0.0048 & 1.2742 & 0.0012 \\ -0.0010 & 0.0004 & -0.3562 & -0.0033 & 0.0012 & 1.2721 \end{bmatrix},$$

$$H = \begin{bmatrix} 0.7346 & -0.0027 & -0.0018 & -0.2057 & 0.0007 & 0.0006 \\ -0.0027 & 0.7329 & 0.0007 & 0.0007 & -0.2052 & -0.0002 \\ -0.0018 & 0.0007 & 0.7318 & 0.0005 & -0.0002 & -0.2049 \\ -0.2057 & 0.0007 & 0.0005 & 0.8406 & 0.0027 & 0.0019 \\ 0.0007 & -0.2052 & -0.0002 & 0.0027 & 0.8423 & -0.0007 \\ 0.0006 & -0.0002 & -0.2049 & 0.0019 & -0.0007 & 0.8435 \end{bmatrix},$$

$$W = \begin{bmatrix} -217.5592 & 0.7428 & -0.5725 & -603.5876 & 123.1579 & 123.7787 \\ -1.0199 & -218.1600 & 0.3047 & 123.2236 & -603.8824 & 123.6221 \\ -0.5871 & 0.3145 & -218.5594 & 123.2847 & 123.7681 & -603.9867 \end{bmatrix}.$$

$$\eta = 1.8807.$$

Then, the controller gain matrix K can be obtained as

$$K = QH^{-1} = \begin{bmatrix} -534.1135 & 42.7502 & 43.1158 & -849.7152 & 160.1318 & 159.6025 \\ 42.2767 & -535.0346 & 44.4461 & 159.8061 & -847.7123 & 156.1346 \\ 42.9813 & 44.4588 & -535.6834 & 158.8171 & 156.4078 & -846.4539 \end{bmatrix}.$$

Case 2. Lyapunov asymptotic stabilization.

By Algorithm 2, we can obtain the qualified feasible set of Problem 2 below

$$P = \begin{bmatrix} 3.0361 & -0.0086 & -0.0086 & 1.2422 & 0.0585 & 0.0513 \\ -0.0086 & 3.0374 & -0.0084 & 0.0434 & 1.2404 & 0.0500 \\ -0.0086 & -0.0084 & 3.0372 & 0.0506 & 0.0514 & 1.2412 \\ 1.2422 & 0.0434 & 0.0506 & 9.7036 & 0.9601 & 0.9577 \\ 0.0585 & 1.2404 & 0.0514 & 0.9601 & 9.7123 & 0.9645 \\ 0.0513 & 0.0500 & 1.2412 & 0.9577 & 0.9645 & 9.7090 \end{bmatrix},$$

$$H = \begin{bmatrix} 0.3477 & 0.0008 & 0.0008 & -0.0449 & 0.0020 & 0.0023 \\ 0.0008 & 0.3475 & 0.0007 & 0.0026 & -0.0449 & 0.0023 \\ 0.0008 & 0.0007 & 0.3475 & 0.0023 & 0.0023 & -0.0449 \\ -0.0449 & 0.0026 & 0.0023 & 0.1108 & -0.0100 & -0.0100 \\ 0.0020 & -0.0449 & 0.0023 & -0.0100 & 0.1107 & -0.0101 \\ 0.0023 & 0.0023 & -0.0449 & -0.0100 & -0.0101 & 0.1107 \end{bmatrix},$$

$$Q = \begin{bmatrix} -10.0000 & 1.8545 & 1.8133 & -10.0000 & 0.5097 & 0.5140 \\ 1.7942 & -10.0000 & 1.8326 & 0.5097 & -10.0000 & 0.5104 \\ 1.8145 & 1.8309 & -10.0000 & 0.5140 & 0.5101 & -10.0000 \end{bmatrix},$$

from which it yields that

$$K = QH^{-1} = \begin{bmatrix} -42.7585 & 5.9275 & 5.7365 & -108.3038 & -2.3458 & -2.2657 \\ 5.5921 & -42.7602 & 5.7790 & -2.2787 & -108.3461 & -2.3347 \\ 5.7346 & 5.7838 & -42.7628 & -2.2725 & -2.3341 & -108.3334 \end{bmatrix}.$$

Case 3. LQR control

Through numerous simulations in trial and error, the following matrices $\mathcal{Q}$ and $\mathcal{R}$, by which a great convergence performance of system (2) can be achieved, are set.

$$\mathcal{Q} = \begin{bmatrix} 12 & 0 & 0 & 0 & 0 & 0 \\ 0 & 16 & 0 & 0 & 0 & 0 \\ 0 & 0 & 20 & 0 & 0 & 0 \\ 0 & 0 & 0 & 12 & 0 & 0 \\ 0 & 0 & 0 & 0 & 20 & 0 \\ 0 & 0 & 0 & 0 & 0 & 14 \end{bmatrix}, \mathcal{R} = \begin{bmatrix} 0.001 & 0 & 0 \\ 0 & 0.002 & 0 \\ 0 & 0 & 0.015 \end{bmatrix}.$$

Then, we can obtain a qualified solution P and a corresponding controller gain K as below

$$P = \begin{bmatrix} 30.5407 & 0.0042 & 0 & 32.8636 & 0.1238 & 0 \\ 0.0042 & 45.1365 & 0 & -0.1011 & 53.6654 & 0 \\ 0 & 0 & 82.7805 & 0 & 0 & 164.3151 \\ 32.8636 & -0.1011 & 0 & 83.6389 & -0.0726 & 0 \\ 0.1238 & 53.6654 & 0 & -0.0726 & 151.3918 & 0 \\ 0 & 0 & 164.3151 & 0 & 0 & 680.1108 \end{bmatrix},$$

$$K = \begin{bmatrix} 109.5452 & -0.3371 & 0 & 278.7964 & -0.2421 & 0 \\ 0.2064 & 89.4424 & 0 & -0.1211 & 252.3197 & 0 \\ 0 & 0 & 36.5145 & 0 & 0 & 151.1357 \end{bmatrix}.$$

Remark 5. *Note that the best results obtained through numerous experiments in trial and error were chosen to be compared to ensure the fairness of the comparison in the above three cases.*

Furthermore, the illustration of the trajectory $x^T(t)Rx(t)$ of the designed spacecraft rendezvous system in the cases of finite-time contractive stabilization, Lyapunov asymptotic stabilization, and LQR control are shown in Figure 2, where $x^T(t)Rx(t)$-FTCS, $x^T(t)Rx(t)$-LAS, and $x^T(t)Rx(t)$-LQR denote the trajectory of $x^T(t)Rx(t)$ under the finite-time contractive stabilization, the Lyapunov asymptotic stabilization, and LQR control, respectively.

In Figure 2, the curve "$x^T(t)Rx(t)$-FTCS" indicates that for given $x(0) = [750, 650, 550, 0, 0, 0]^T$, which satisfies $x^T(0)Rx(0) = 1.2875 \times 10^6 < 1.3 \times 10^6$, $x^T(t)Rx(t) < 2.5 \times 10^6$

holds, $\forall\, t \in [0, 40]$, $x^T(t)Rx(t) < 1.0 \times 10^4$ holds, $\forall\, t \in [10, 40]$. Hence, according to Definition 1, the designed spacecraft rendezvous system is finite-time contractively stable w.r.t. $(1.3 \times 10^6, 2.5 \times 10^6, 1.0 \times 10^4, I, 10, 40)$ under finite-time contractive stabilization. In addition, the curve "$x^T(t)Rx(t)$-LAS" shows that the trajectory of $x^T(t)Rx(t)$ reaches the bound c_3 at 7.947 s for the first time under the case of Lyapunov asymptotic stabilization; however, it escapes from the bound at the time interval [10, 14.022]. Hence, the designed spacecraft rendezvous system is Lyapunov asymptotically stable but is not finite-time contractively stable w.r.t. $(1.3 \times 10^6, 2.5 \times 10^6, 1.0 \times 10^4, I, 10, 40)$ in this case, which verifies the conclusion that if a system is finite-time contractively stable, it must be Lyapunov asymptotically stable, but not vice versa.

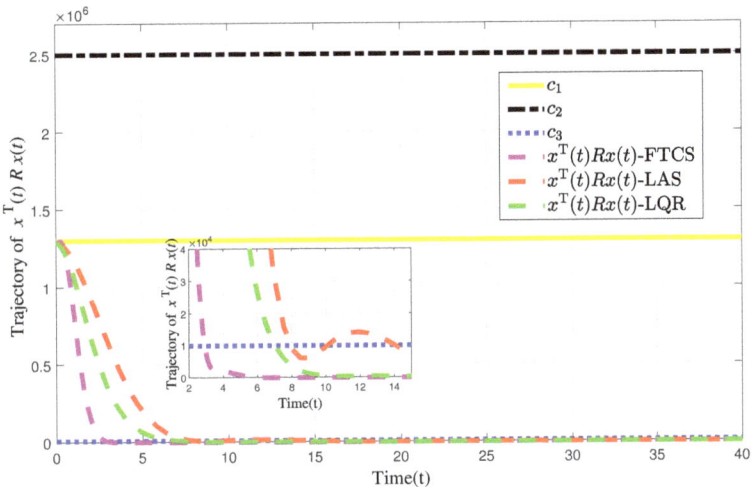

Figure 2. Evolution of $x^T(t)Rx(t)$ under different control methods.

The trajectories of relative position and velocity under different control laws are presented in Figures 3–8. Moreover, the distribution diagram of pole points of the resulting closed-loop system from input u_1 to all outputs (y_1, y_2, y_3) is shown in Figure 9. (Such diagrams from u_2, u_3 to all outputs (y_1, y_2, y_3) are the same as that from u_1 to all outputs in this case. Here, they are not listed for simplicity.) It follows from Figures 3–8 that the relative position and velocity of the considered two spacecrafts along the x, y, and z axes gradually reduce to 0 under cases of finite-time contractive stabilization and Lyapunov asymptotic stabilization. This indicates that the spacecraft rendezvous can be achieved through the use of the designed controllers. Furthermore, comparing $x(t)$-FTCS, ..., $\dot{z}(t)$-FTCS with $x(t)$-LAS, ..., $\dot{z}(t)$-LAS and $x(t)$-LQR, ..., $\dot{z}(t)$-LQR, respectively, it is attained that in the case of finite-time contractive stabilization, the achievement of the spacecraft rendezvous is quicker than that in the case of Lyapunov asymptotic stabilization and LQR control, from which it can be concluded that the convergence performance for finite-time contractive stability can be better than Lyapunov asymptotic stability. This conclusion can also be supported intuitively by Figure 9, where poles in the FTCS case that all lie on the left of "$s = -\alpha$" are definitely farther from the imaginary axis than that in the LAS and LQR cases.

In addition, according to the simulation results, if assuming that the chaser spacecraft needs to approach the target spacecraft within a short enough t_s, there is no doubt that the strength of thrust of the chaser spacecraft is suffering challenges in the consideration of finite-time contractive stabilization in this paper, and thus, the corresponding cased energy consumption has to be accommodated. Actually, a balance between the expected t_s and acceptable strength of thrust is needed in practice.

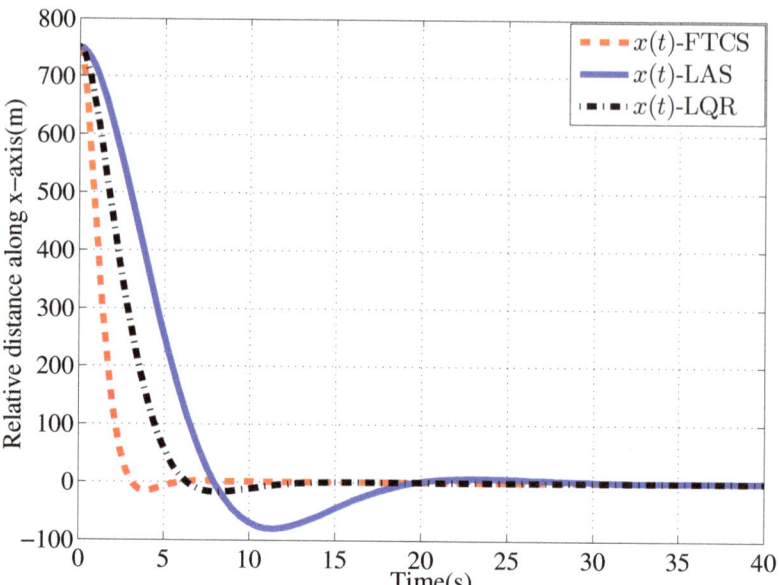

Figure 3. Relative position along x-axis under different control laws.

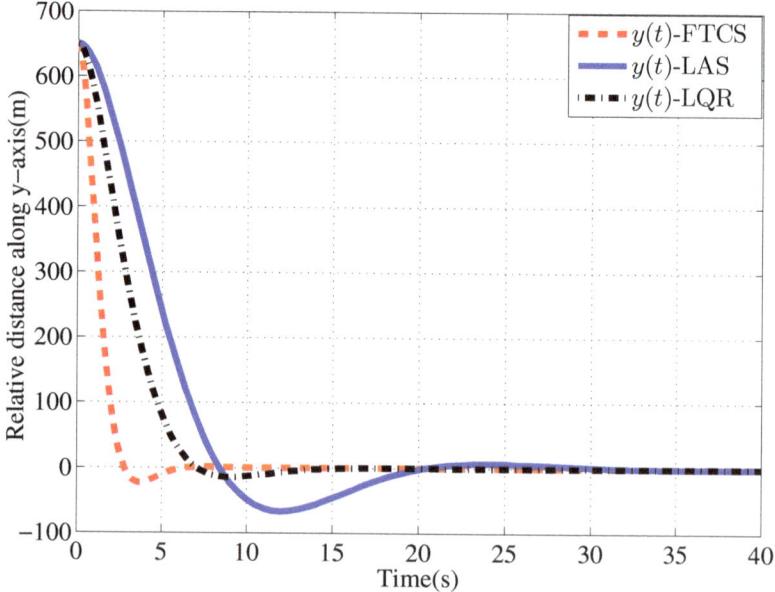

Figure 4. Relative position along y-axis under different control laws.

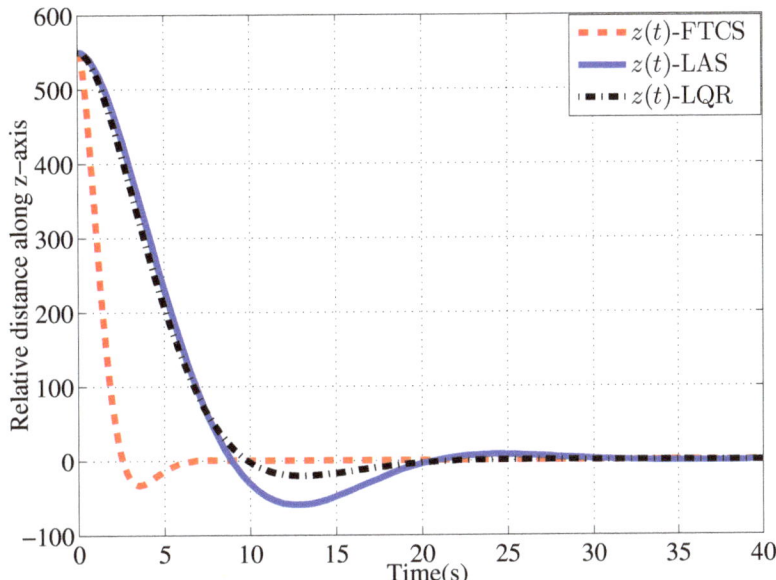

Figure 5. Relative position along z-axis under different control laws.

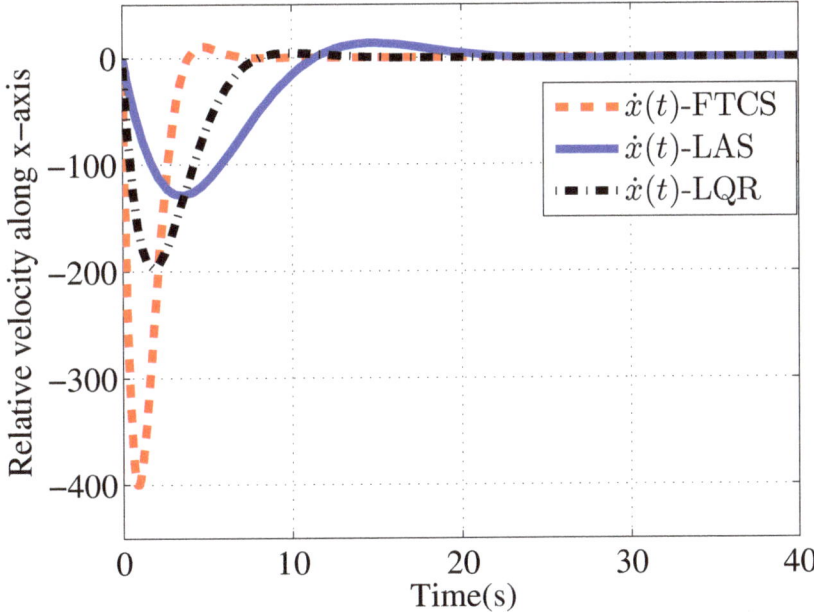

Figure 6. Relative velocity along x-axis under different control laws.

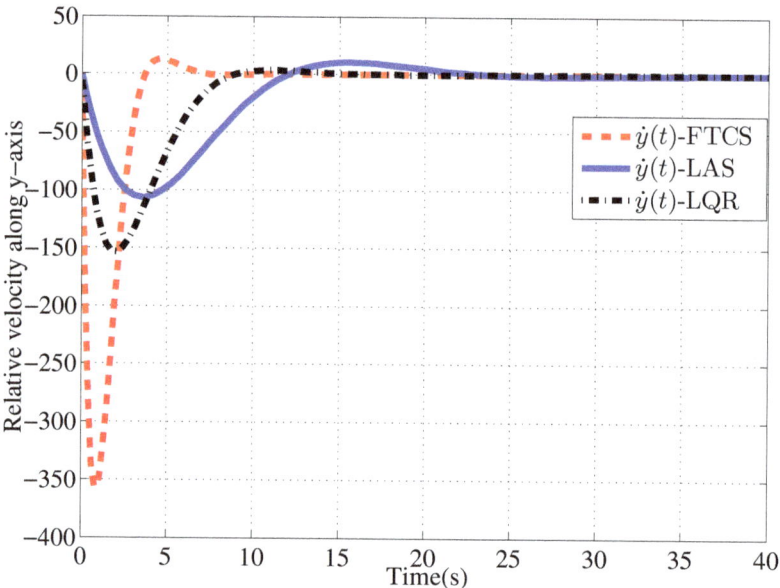

Figure 7. Relative velocity along y-axis under different control laws.

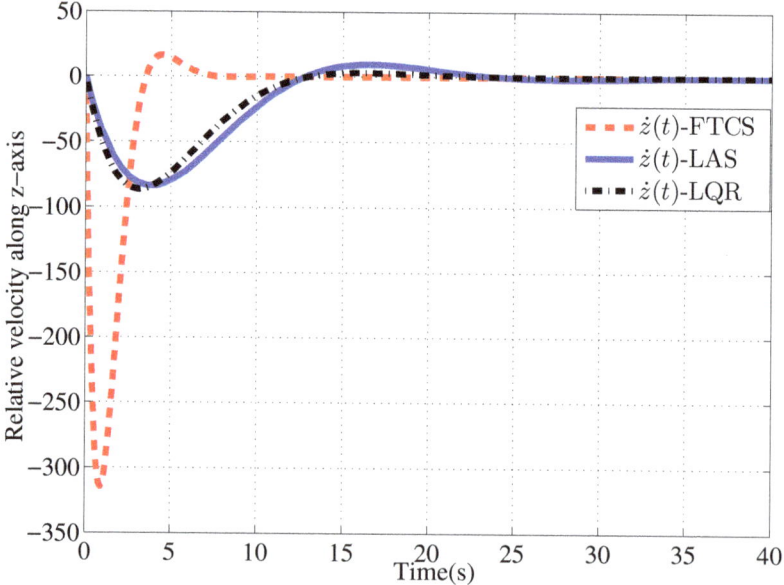

Figure 8. Relative velocity along z-axis under different control laws.

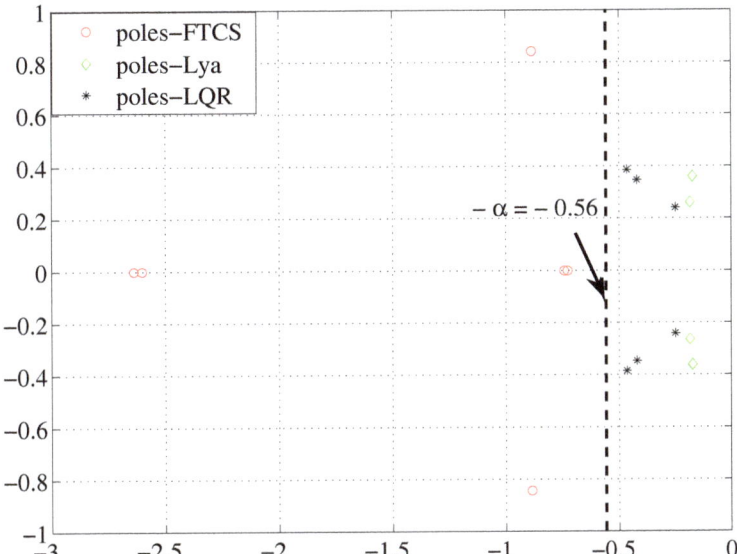

Figure 9. Distribution diagram of pole points from input u_1 to all outputs.

5. Conclusions

The finite-time contractive control problem for spacecraft rendezvous was investigated in this paper. Based on the Lyapunov stability theory, the existence condition of the finite-time contractive controller was established. The cone complementary linearization technique was adopted to make the controller design numerically tractable. An illustrative example showed the effectiveness of the proposed controller. Considering that the state unavailability and noise of the spacecraft rendezvous system, which we ignored in this paper, commonly need to be considered in practice, the future research interests of this paper include the robust finite-time contractive boundedness control issue for an uncertain spacecraft rendezvous system with disturbance and noise effects under observer-based dynamic output feedback control, and the guaranteed cost finite-time contractive stabilization issue for an uncertain spacecraft rendezvous system with/without disturbances and noise effects.

Author Contributions: Methodology, M.L.; Formal analysis, B.Z.; Writing—original draft, J.S.; Writing—review & editing, Y.G. All authors have read and agreed to the published version of the manuscript.

Funding: This research was funded by National Natural Science Foundation of China grant number 11972130.

Conflicts of Interest: The authors declare no conflict of interest.

References

1. Yamanaka, K.; Ankersen, F. New state transition matrix for relative motion on an arbitrary elliptical orbit. *J. Guid. Control Dyn.* **2002**, *25*, 60–66. [CrossRef]
2. Clohessy, W.H.; Wiltshire, R.S. Terminal guidance system for satellite rendezvous. *J. Aerosp. Sci.* **1960**, *27*, 653–658. [CrossRef]
3. Tschauner, J.; Hempel, P. Rendezvous Zueinem in Elliptischer Bahn Umlaufenden Ziel. *Astronaut. Acta* **1965**, *11*, 104–109.
4. Sheng, D.; Yang, X.; Karimi, H.R. Robust control for autonomous spacecraft evacuation with model uncertainty and upper bound of performance with constraints. *Math. Prob. Eng.* **2014**, *2014*, 589381. [CrossRef]
5. Yang, X.; Gao, H. Guaranteed cost output tracking control for autonomous homing phase of spacecraft rendezvous. *J. Aerosp. Eng.* **2011**, *24*, 478–487. [CrossRef]
6. Guo, Y.; Zhang, D.; Li, A. Finite-time control for autonomous rendezvous and docking under safe constraint. *Aerosp. Sci. Technol.* **2021**, *109*, 106380. [CrossRef]

7. Luo, W.; Zhou, B.; He, L.; Duan, G. Global stabilization of the spacecraft rendezvous system by delayed and bounded linear feedback. *IEEE Trans. Syst. Man Cybern. Syst.* **2020**, *52*, 1373–1384. [CrossRef]
8. Zhang, K.; Liu, Y.; Tan, J. Semi-global finite-time stabilization of saturated spacecraft rendezvous system by dynamic event triggered and self-triggered control. *IEEE Trans. Aerosp. Electron. Syst.* **2022**, *58*, 5030–5042. [CrossRef]
9. Wang, Q.; Xue, A. Robust control for spacecraft rendezvous system with actuator unsymmetrical saturation: A gain scheduling approach. *Int. J. Control* **2018**, *91*, 1241–1250. [CrossRef]
10. Huang, C.; Cao, T.; Huang, J. Autonomous Control of the Large-Angle Spacecraft Maneuvers in a Non-Cooperative Mission. *Sensors* **2022**, *22*, 8586. [CrossRef]
11. Weiss, L.; Infante, E. Finite time stability under perturbing forces and on product spaces. *IEEE Trans. Autom. Control* **1967**, *12*, 54–59. [CrossRef]
12. Onori, S.; Dorato, P.; Galeani, S. Finite time stability design via feedback linearization. In Proceedings of the 44th IEEE Conference on Decision and Control, Seville, Spain, 10–13 December 2005; pp. 4915–4920. [CrossRef]
13. Li, X.; Yang, X.; Song, S. Lyapunov conditions for finite-time stability of time-varying time-delay systems. *Automatica* **2019**, *103*, 135–140. [CrossRef]
14. Cheng, J.; Xiang, H.; Wang, H. Finite-time stochastic contractive boundedness of Markovian jump systems subject to input constraints. *ISA Trans.* **2016**, *60*, 74–81. [CrossRef] [PubMed]
15. Yang, X.; Li, X. Finite-time stability of nonlinear impulsive systems with applications to neural networks. *IEEE Trans. Neural Netw. Learn. Syst.* **2021**, *34*, 243–251. [CrossRef]
16. Zhang, X.; Li, C.; Li, H. Finite-time stabilization of nonlinear systems via impulsive control with state-dependent delay. *J. Frankl. Inst.* **2022**, *359*, 1196–1214. [CrossRef]
17. Wang, Z.; Sun, J.; Chen, J. Finite-time stability of switched nonlinear time-delay systems. *Int. J. Robust Nonlinear Control* **2020**, *30*, 2906–2919. [CrossRef]
18. Zhang, T.; Li, X.; Song, S. Finite-Time Stabilization of Switched Systems Under Mode-Dependent Event-Triggered Impulsive Control. *IEEE Trans. Syst. Man Cybern. Syst.* **2021**, *52*, 5434–5442. [CrossRef]
19. Joby, M.; Santra, S.; Anthoni, S.M. Finite-time contractive boundedness of extracorporeal blood circulation process. *Appl. Math. Comput.* **2021**, *388*, 125527. [CrossRef]
20. Zhang, H.; Georgescu, P. Finite-time control of impulsive hybrid dynamical systems in pest management. *Math. Methods Appl. Sci.* **2014**, *37*, 2728–2738. [CrossRef]
21. Yang, X.; Cao, X.; Gao, H. Sampled-data control for relative position holding of spacecraft rendezvous with thrust nonlinearity. *IEEE Trans. Ind. Electron.* **2011**, *59*, 1146–1153. [CrossRef]
22. Song, X.; Zhou, S.; Zhang, B. A cone complementarity linearization approach to robust $H\infty$ controller design for continuous-time piecewise linear systems with linear fractional uncertainties. *Nonlinear Anal. Hybrid Syst.* **2008**, *2*, 1264–1274. [CrossRef]
23. Michel, A.; Porter, D. Practical stability and finite-time stability of discontinuous systems. *IEEE Trans. Circuit Theory* **1972**, *19*, 123–129. [CrossRef]
24. Dorato, P. Short time stability in linear time-varying systems. In Proceedings of the IRE International Convention Record, New York, NY, USA, 20–23 March 1961; Part 4, pp. 83–87.
25. Amato, F.; Ambrosino, R.; Ariola, M.; Cosentino, C.; De Tommasi, G. *Finite-Time Stability and Control*; Springer: Berlin/Heidelberg, Germany, 2014. [CrossRef]
26. El Ghaoui, L.; Oustry, F.; AitRami, M. A cone complementarity linearization algorithm for static output-feedback and related problems. *IEEE Trans. Autom. Control* **1997**, *42*, 1171–1176. [CrossRef]
27. Dorato, P.; Cerone, V.; Abdallah, C. *Linear-Quadratic Control: An Introduction*; Simon & Schuster, Inc.: New York, NY, USA, 1994.
28. Löfberg, J. YALMIP: A Toolbox for Modeling and Optimization in MATLAB. *IEEE Int. Symp. Comput. Aided Control Syst. Des.* **2004**, *3*, 282–289. [CrossRef]

Disclaimer/Publisher's Note: The statements, opinions and data contained in all publications are solely those of the individual author(s) and contributor(s) and not of MDPI and/or the editor(s). MDPI and/or the editor(s) disclaim responsibility for any injury to people or property resulting from any ideas, methods, instructions or products referred to in the content.

Three-Dimensional Path Planning of UAV Based on Improved Particle Swarm Optimization

Lixia Deng *, Huanyu Chen, Xiaoyiqun Zhang and Haiying Liu

School of Information and Automation Engineering, Qilu University of Technology (Shandong Academy of Sciences), Jinan 250353, China
* Correspondence: lixiadeng@qlu.edu.cn

Abstract: The traditional particle swarm optimization algorithm is fast and efficient, but it is easy to fall into a local optimum. An improved PSO algorithm is proposed and applied in 3D path planning of UAV to solve the problem. Improvement methods are described as follows: combining PSO algorithm with genetic algorithm (GA), setting dynamic inertia weight, adding sigmoid function to improve the crossover and mutation probability of genetic algorithm, and changing the selection method. The simulation results show that the improved PSO algorithm solves better route results and is faster and more stable.

Keywords: particle swarm algorithm; UAV; 3D path planning; SHADE algorithm

MSC: 93C85

1. Introduction

Currently, as robots enter our lives, boring, repetitive work is being transformed to a more unmanned and intelligent system using machines instead of work. Among them, the development of technologies related to drones has brought great convenience to our lives, such as inspecting and exploring dangerous environments, delivering deliveries, power patrols, and other tasks. For performing complex tasks in low-altitude flight, UAV navigation capabilities and planning paths are particularly important [1].

In 1959, Dr. Dantzig and Dr. Ramser first raised the vehicle-based routing problem. From then on, the routing problem has become a new research topic for domestic and foreign scholars [2]. Path planning algorithms suitable for UAV use can be divided into two categories, global path planning algorithms in the continuous domain range and local path planning algorithms in the continuous domain range [3]. In a two-dimensional environment, traditional algorithms such as A* algorithm [4], Dijkstra algorithm [5], and simulated annealing algorithm [6] perform search to find the optimal path by means of a raster map. However, when extended to three dimensions, the consumption of time and memory increases proportionally and is no longer applicable with path planning.

Sampling-based stochastic methods [7] have received wide application and attention in recent years. Random sampling in state space can effectively solve high-dimensional and complex path planning problems, but there are problems of low accuracy and slow convergence [8].

Inspired by nature, swarm intelligence algorithms are applied to route planning in complex environments, for example, particle swarm algorithm [9], ant colony algorithm [10], and genetic algorithm [11]. These algorithms find an efficient path through different strategies and then obtain the final path by continuous iterations. However, the tendency to fall into local optimum is still one of the problems that swarm intelligence algorithms need to solve.

Three-dimensional path planning is global path planning, which finds the best and collision-free path in 3D clutter considering geometric, physical, and temporal constraints [12].

This requires high accuracy and speed of the algorithm. The algorithms presented above have been applied to 3D path planning, among which the swarm intelligence algorithm stands out due to its excellent search capability. However, a single swarm intelligence algorithm still has limitations, and how to improve it has become one of the topics of scholars Wang Yihu et al. [13] introduced the convergence and migration operations of the bacterial foraging algorithm (BFO) in the PSO algorithm, which effectively improved some of the defects of the PSO algorithm and improved its search capability. Xueying Sun et al. [14] proposed a high-performance bacterial foraging-genetic-particle swarm hybrid algorithm to address the defects of the particle swarm algorithm, which improved the computational speed and capability of the algorithm and further improved the usability of the method. B Abhishek et al. [15] proposed a harmony-based search algorithm, which performs both exploratory search and usage search, and further optimizes the generated path combined with unmanned aerial vehicle constraints, while speeding up the algorithm to avoid falling into local optimum. Manh Duong Phung [16] presents a spherical quantum-oriented particle swarm optimization algorithm (SPSO), which transforms the path planning problem into an optimization problem containing the requirements and constraints for UAV's feasible and safe operation. The SPSO algorithm is used to find the optimal path through the relationship between particle position and UAV speed, turn angle, and pitch angle.

In this paper, the hybrid particle swarm optimization algorithm CPSO is designed to solve the UAV 3D path planning problem by combining the excellent design.

- Establish the experimental environment model, set 3D mountains as obstacles, construct fitness function based on obstacles and path length, and introduce cubic B-spline curve to smooth the path.
- Set the adaptive dynamic inertia weight of the adaptive particle swarm optimization algorithm to ensure the early search ability while enhancing the optimization ability of the later population.
- By introducing the selection operation of improved SHADE algorithm and improved genetic algorithm, the population diversity is improved, avoiding falling into local optima and reducing search time.
- Finally, comparative simulation experiments were conducted using MATLAB to compare with various swarm intelligence algorithms, including particle swarm optimization (PSO), particle swarm optimization algorithm with adaptive inertia weights (wPSO), SHADE algorithm (SHADE), genetic algorithm (GA), and ant colony algorithm (ACA), taking into account the overall performance of the algorithm.

The other parts of this article are arranged as follows: Section 2 establishes the experimental environment and path smoothing algorithm; Section 3 introduces the improvement strategy of CPSO in detail; Section 4 introduces the experimental environment, experimental results analysis, and algorithm comparison. Finally, the conclusion is given in Section 5, and the focus and improvement direction of follow-up work are proposed.

2. Model Establishment

This section describes the environmental model and path smoothing model used in this experiment, specifically for building a three-dimensional environmental model with mountain barriers and a cubic B-spline curve for smooth paths.

2.1. Environmental Model

The 3D path planning of the UAV needs to obtain information from the terrain model, and the actual situation should be considered when modeling the terrain. By considering obstacles, environment, and other factors, the established terrain model [17] is described as follows:

$$Z_1(x,y) = \sin(y+a) + b \cdot \sin(x) + c \cdot \cos(d \cdot \sqrt{x^2+y^2}) \\ + e \cdot \cos(y) + f \cdot \sin(g \cdot \sqrt{x^2+y^2}) \tag{1}$$

where x and y are the horizontal coordinates, and Z_i are the corresponding height values. a, b, c, d, e, f, and g are constant coefficients that control the undulation of the base terrain and can be set as needed or generated randomly. For a mountain in 3D environment, it can be represented by the following model:

$$z(x,y) = \sum_{i=1}^{P} h_i \exp\left[-\left(\frac{x-x_i}{x_{si}}\right)^2 - \left(\frac{y-y_i}{y_{si}}\right)^2\right] \qquad (2)$$

where P represents the total number of mountain peaks, (x_i, y_i) represents the center coordinate of the i-th peak, and h_i is the parameter that controls the height. x_{si} and y_{si} are the attenuations of the i-th peak along the x-axis and y-axis which can be used to control the slope, respectively.

2.2. Path Smoothing Algorithm Based on Cubic B-Spline Curve

In order to prevent frequent angle adjustment during the flight, ensure the safety of the UAV, and reduce the sailing time, a cubic B-spline curve is introduced [18]. In a given $m + n + 1$ plane or space vertex $P_i(i = 0, 1, \ldots, m+n)$, it is called a parametric curve segment of degree n:

$$P_{k,n}(t) = \sum_{i=0}^{n} P_{i+k} G_{i,n}(t) \, t \in [0,1] \qquad (3)$$

where $P_{k,n}(t)$ is the n-th degree B curve segment of the k-th segment, and these curve segments are called n-th degree B-spline curves. $G_{i,n}(t)$ is the basis function which is defined based on Equation (4).

$$G_{i,n}(t) = \frac{1}{n!} \sum_{j=0}^{n-i} (-1)^j C_{n+1}^j (t+n-i-j)^n \qquad (4)$$

$$t \in [0,1] \, i = 0, 1, \ldots n$$

in order to ensure the smoothness of the path and consider the difficulty, let $n = 3$, and a cubic B-spline curve is used to smooth the path.

3. Improved Particle Swarm CPSO Algorithm Design

This section describes the design ideas of the improved particle swarm optimization algorithm CPSO, including constructing a fitness function based on barriers and path length, changing fixed weights to adaptive dynamic weights to improve the optimization ability, and finally fusing the SHADE algorithm to improve population diversity.

3.1. Particle Swarm Algorithm

Particle swarm optimization (PSO) is an evolutionary computing technique. Inspired by the results of artificial life studies, Dr. Eberhart and Dr. Kennedy proposed a particle swarm algorithm by simulating bird foraging migration and population behavior and improving Craig Reynolds' bird cluster model [19]. A massless particle is designed to simulate a bird in a flock, and the particle has only two attributes: speed and position, where speed represents the speed of movement and location represents the direction of movement. Each particle individually searches for the optimal solution, records it as the current individual extreme value, and shares the individual extreme value with other particles in the whole particle swarm to find the optimal individual extreme value as the current global optimal solution for the whole particle swarm. All particles in the particle swarm adjust their speed and position according to the current individual extreme value they find and the current global optimal solution shared by the whole particle swarm. The basic idea of the particle swarm optimization algorithm is to find the best solution through collaboration and information sharing among individuals in a group. The particle swarm optimization algorithm operates on particles using the following formulas:

$$v_{ij}(t+1) = wv_{ij}(t) + c_1 r_1(t)[pbest_i - x_{ij}(t)] + c_2 r_2(t)[gbest_i - x_{ij}(t)] \qquad (5)$$

$$x_{ij}(t+1) = x_{ij}(t) + v_{ij}(t+1) \qquad (6)$$

where w denotes the inertial weight and the degree of trust in the current speed direction c_1, c_2 is the learning factor, also known as the acceleration constant; r_1, r_2 is a random value between 0 and 1, increasing search randomness. v_{ij} is the velocity of the particle. $pbest$ is the best position for the i particle to experience, and $gbest$ is the best position for all particles of the group to experience. x_{ij} is the current position of the particle.

3.2. Fitness Function Design

The quality of the path length is one of the important indicators to measure the success of the algorithm improvement. Due to the lack of battery capacity of the UAV, the flight distance is limited. The shorter the flight path, the less time and energy it takes.

Based on the cubic B-spline curve fitting path, the interpolation process is performed, and the interpolation is differentiated to obtain the fitness function:

$$fitness = \sqrt{(x_{i+1} - x_i)^2 + (y_{i+1} - y_i)^2 + (z_{i+1} - z_i)^2} \qquad (7)$$

where (x_i, y_i, z_i) are the coordinates of the i node of the path; $(x_{i+1}, y_{i+1}, z_{i+1})$ are the coordinates of the $i + 1$ node. The obstacle risk factor f is introduced to avoid the collision between the UAV and the obstacle. The barrier coefficient formula is described as follows:

$$f = \begin{cases} 0 & L_{min} > L_d \\ 1 & L_{min} < L_d \end{cases} \qquad (8)$$

considering the real environment, UAV is not a particle and it has its own size. So, L_{min} is set as the minimum distance close to the peak, and L_d as the safe distance. Combined with laboratory drone data, set $L_d = 0.12$. When $f = 1$, the minimum distance is less than the safe distance, which is prone to danger, so it is necessary to increase the fitness function. The fitness function is changed to:

$$fitness = kfitness \qquad (9)$$

where k is the multiple of expansion, which can be set according to the experimental environment. This experiment has set $k = 5$.

3.3. Adaptive Dynamic Inertial Weight

The inertia weight is an important control parameter in the particle swarm algorithm, and the size of the inertia weight indicates how much of the current velocity inheritance goes to the particle. If inertia weight is set larger, the global search ability is stronger, and if inertia weight is set smaller, the local search ability is stronger and the global search ability becomes weaker [20]. In this paper, a linearly decreasing inertia weight is designed. In the early stage of the algorithm, a larger inertia weight is used to ensure the global search ability. With the increase in number of iterations, the inertia weight becomes smaller and the local search ability is enhanced. The formula is described as follows:

$$w = (w_{max} - w_{min}) \times \frac{(N - iter)}{N} + \frac{(w_{max} - w_{min})}{iter} \qquad (10)$$

where $w_{\max}$ is the maximum inertia weight, $w_{\min}$ is the minimum inertia weight, N is the maximum number of iterations, and $iter$ is the current iteration number of the algorithm.

3.4. Improved SHADE Algorithm

Differential evolution algorithm (DE) [21] belongs to one of the evolutionary algorithms (EA). It includes the following steps: 1. initialization of the population. 2. mutation operation. 3. crossover operation. 4. selection operation. Then after that, the JADE [22] algorithm is updated, which has the same logic as DE. Subsequently, SHADE [23] was introduced on the basis of JADE. In this paper, the DE family algorithm is chosen to increase the population diversity, prevent falling into local optimum, and at the same time improve the operation speed.

According to the characteristics of DE, JADE, and SHADE algorithms, the variation operation and crossover operation are taken to be used in the algorithm. The number of populations in this experiment is M = 50; if each iteration does not consider the interference of invalid points, this will lead to an increase in the algorithm running speed. By the selection operation of hybrid genetic algorithm GA, the good individuals are left and the bad ones are eliminated.

This paper adopts the mixed selection operator. The first method uses the optimal fitness selection method to sort the fitness and selects the better fitness as the parent 1, with selection of populations at 50%. The second method uses the roulette method by selecting the probability $p_{\sec}$. The selected population is used as parent 2, and the population share is 50%. The combination becomes the parent of the next iteration.

The variation operation achieves individual variation through a difference strategy, and the improved variation strategy is chosen in this paper: DE/current-to-best/1. The equation is as follows:

$$V_{i,j,v}(g+1) = X_{i,G} + F_i \cdot \left(X_{best,G}^P - X_{i,G}\right) + F_i \cdot (X_{r_1^i,G} - X_{r_2^i,G}) \tag{11}$$

$$\mu_F = (1-c) \cdot \mu_F + c \cdot mean(S_F) \tag{12}$$

$$F_i = rand(\mu_F, 0.1) \tag{13}$$

where $X_{i,G}$ is the i particle being processed, $X_{best,G}^P$ is an individual in the top $p*M$ of the current population fitness ranking, p is a given proportion, p = 10%, $X_{r_1^i,G}$ is the i particle being processed of parent 1 selected by GA, $X_{r_2^i,G}$ is the i-th particle being processed of parent 2 selected by GA, $\mu_F = 0.4$, F_i is the scale factor, $c = \frac{1}{10}$, and $mean(S_F)$ is the ratio of parent optimal fitness function to population size.

Afterwards, crossover operations between individuals were performed to improve population diversity for the populations after mutation operations. The formula is as follows:

$$u_{j,i,v}(g+1) = \begin{cases} v_{j,i,v}(g+1), if rand(0,1) \leq C_r \\ x_{j,i,v}(g+1), otherwise \end{cases} \tag{14}$$

$$\mu_{CR} = (1-c) \cdot \mu_{CR} + c \cdot mean(S_{cr}) \tag{15}$$

$$C_r = rand(\mu_{CR}, 0.1) \tag{16}$$

where $x_{j,i,v}(g+1)$ is the parent population, C_r is the crossover probability, $\mu_{CR} = 0.5$, $c = \frac{1}{10}$, and $mean(S_{cr})$ is the ratio of the current population optimal fitness function to the population size.

3.5. Constraint Condition

In order to prevent the UAV from being dangerous during flight, constraints need to be set according to the actual situation. Firstly, the altitude has a great influence on the UAV. Flight at high altitude is susceptible to temperature and airflow, and flight at low altitudes is susceptible to disturbances from buildings and trees. Therefore, flying at the appropriate altitude can be expressed as Equation (13):

$$z_{min} < z_j < z_{max} \qquad (17)$$

where z_j is the height position of the j-th time. z_{min}, z_{max} represent the minimum and maximum heights, and $z_{min} = 5, z_{max} = 100$ is set according to the actual situation. Meanwhile, the size of environment is set to 100 × 100 × 250 m to prevent the UAV from flying out of the set environment.

4. Experimental Simulation and Analysis

Firstly, summarize the algorithm flowchart according to the third section, and then introduce the experimental hardware configuration and some algorithm parameters of this experiment. Perform comparative analysis of algorithms in the same and random environments to validate the improved particle swarm optimization algorithm.

4.1. Improved Algorithm Operation Flow

The flow chart of the improved PSO algorithm is presented in Figure 1.

(1) Establishing an experimental environment according to Equations (1) and (2); refer to Section 2.1. Setting the start point and end point. The starting point is represented as a box and the ending point is an asterisk.

(2) Parameter initialization. Setting the particle population size, maximum number of iterations, inertia weight, social weight, and cognitive weight. For parameter settings, refer to Table 1.

(3) Population initialization. Randomly generating particles and initializing the velocity, calculating the initial fitness and performing collision detection, and updating the individual optimum as well as the global optimum.

(4) Enter the main loop. Update velocity and position, perform velocity and position detection at the same time to avoid out-of-bounds, calculate fitness values and perform collision detection, update individual optimum and global optimum.

(5) The selection operation of genetic algorithm is introduced to select the outstanding particle population as the parent, followed by the crossover and mutation operation of SHADE algorithm. The new population is used as the initial population for the next performed cycle.

(6) End condition. Determine if the maximum number of iterations has been reached, and if so, exit the loop and output the result; otherwise, return to step (4).

4.2. Experimental Environment and Parameters

To verify the advantages of the proposed CPSO algorithm, traditional PSO algorithm, SHADE, and PSO with modified dynamic inertia weights (wPSO) are used as control group, and the parameters such as the number of iterations remain unchanged. The above three algorithms are simulated and tested on MATLAB. Two sets of experiments are carried out and the experimental results are analyzed. The test environment is Windows 10, 64-bit system, MATLAB R2020b simulation platform. Parameters in the algorithm are shown in Table 1:

In order to verify the superiority of the improved PSO algorithm in 3D path planning, the following two experiments are carried out, respectively.

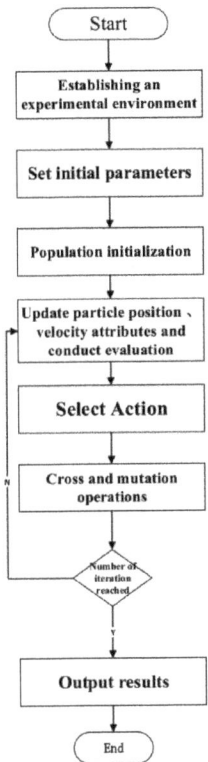

Figure 1. Flow chart.

Table 1. Algorithm parameters.

Quantity	Symbol	Numerical Value
Spatial scope	/	100 × 100 × 250
Starting point	start	(1,1,1)
End point	goal	(95,76,30)
Total group number	M	50
Number of iterations	N	200
Current iterations	iter	/
Social weight	c_1	2
Cognitive weight	c_2	2
Maximum inertia weight	w_{max}	0.9
Minimum inertia weight	w_{max}	0.4
Expand multiple	k	5

4.3. Comparative Analysis in the Same Simple Environment

This section conducted comparative analysis experiments on algorithms in simple and complex environments. Figure 2 shows the front view of 3D path planning results in a simple environment, and Figure 3 shows the front view of 3D path planning results in a complex environment. Figures 4 and 5 show the fitness curves of simple and complex environments, respectively.

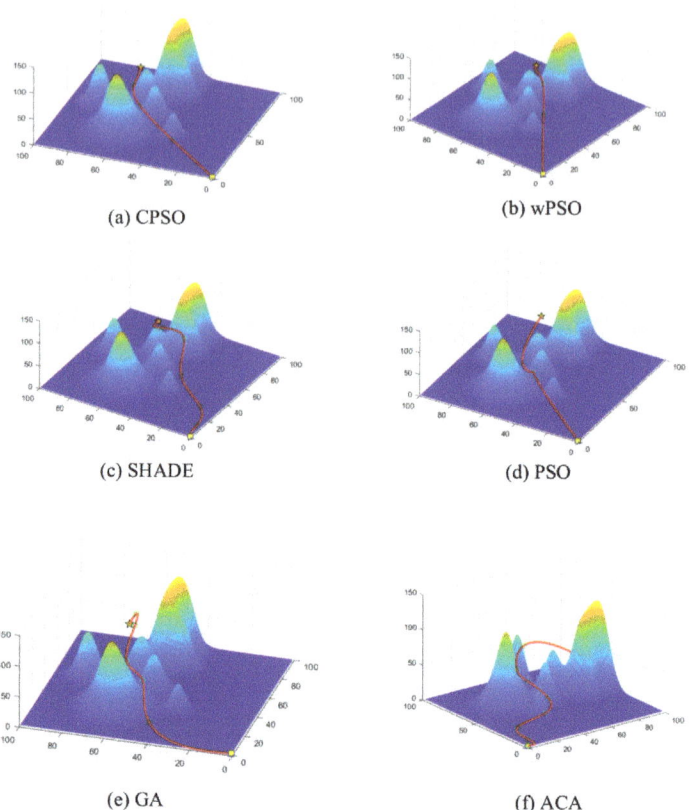

Figure 2. Simple environment result diagram.

From Figures 2 and 3, it can be seen that several algorithms can complete path planning tasks in a three-dimensional environment. From the e and f graphs in Figure 3, it can be observed that the GA and ACA algorithms generate long paths with multiple inflection points, and although they can stay away from obstacles, they are not the optimal choice. Compared to e and f graphs, the path generation of C graph is better, but there is still a significant curve situation. The generation path of D graph is excellent, but the inflection points can be clearly found in the graph. The simple PSO algorithm is not sufficient for path smoothness, so it is not adopted. The path generation in Figure 2b is smooth, but compared to Figure 2a, the degree of path optimization is clearly not excellent enough. Overall, the CPSO generated in Figure 2a has a smooth path, no inflection points, good continuity, and the shortest path.

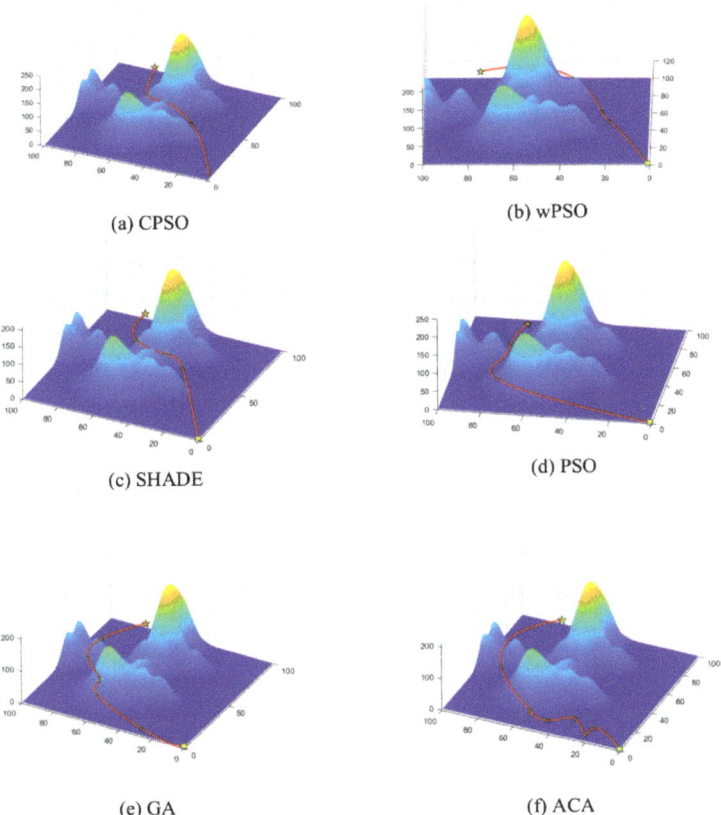

(a) CPSO (b) wPSO (c) SHADE (d) PSO (e) GA (f) ACA

Figure 3. Complex environment result diagram.

From the fitness curve in Figure 6, it can be seen that the PSO algorithm and wPSO algorithm have been trapped in local optimum around 20 iterations, lacking the ability to jump out of local optimum, among which PSO algorithm has been trapped in local optimum in the ninth iteration. Although the search was still in progress at the 115th iteration, the effect was no longer apparent. Because of the inertial dynamic weights, wPSO reaches its optimum in about 20 generations and still tries to optimize in later generations, but the diversity of the population is insufficient, resulting in poor search ability in later generations. Because the SHADE algorithm needs to be continuously evolved, it takes a long time to reach its optimal level in about 60 generations, and the performance of the algorithm is not as good as other control groups. GA and ACA algorithms have large fitness and weak optimization ability, which require great computing power for 3D path optimization tasks, so there is still great room for the development of the two algorithms. The CPSO algorithm incorporates dynamic inertia weights, which are larger in the pre-iteration period; it guarantees global searching ability and sharply reduces the fitness curve. As the number of iterations increases, the weight decreases, the local optimization ability is strengthened, and the convergence rate is increased, and the optimal result is basically reached in 23 generations. The selection operation in the genetic algorithm and the crossover and mutation operation in SHADE are introduced to improve the diversity of the population, enhance the search ability, and still perform local optimization in the late iteration.

For the complex path planning in Figure 3, the differences in path generation are particularly prominent. Firstly, the drawbacks of Figure 3e,f in simple obstacles are magnified in complex obstacles, resulting in more inflection points and longer generation paths proving that the two algorithms have poor processing ability in complex three-dimensional environments. The d and b graphs perform well in simple environments, but their own shortcomings are exposed when the environment is complex. Firstly, there is a problem of generating path differences in both Figure 3b,d, and there are large-scale curves, which can lead to the risk of drone crashes during flight. In addition, there are also a few inflection points in the paths generated by the two, further increasing the danger of drone flight. From the image, CPSO and SHADE are both excellent for complex 3D path planning problems, but from the fitness Figure 5, it can be found that CPSO's fitness curve is significantly lower than SHADE's, so the path generated by CPSO is better.

In summary, in increasingly complex environments, CPSO performs significantly better than other comparative algorithms, so CPSO can be used as a choice for dealing with path planning problems in complex environments.

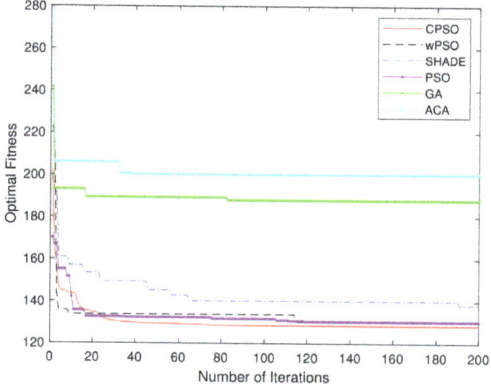

Figure 4. Simple environmental fitness value.

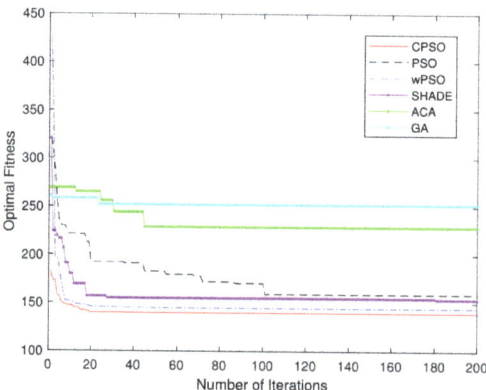

Figure 5. Complex environmental fitness value.

4.4. Comparative Analysis in Random Environment

Establish a three-dimensional random environment within the range of $100 \times 100 \times 100$ m, with a starting point of (1,1,1) m and an endpoint of (100,100,50) m. Randomly generate 10 obstacles with 200 iterations. Perform 10 simulation tests on the 6 algorithms mentioned above, and the simulation results are shown in Figure 6. The statistical results are shown in Table 2. It can be found from the statistical table that the average fitness of CPSO algorithm is significantly lower than that of other comparison algorithms, showing good optimization ability. Through variance comparison, CPSO has good stability. According to Table 2, by comparing the number of iterations and time to reach the average fitness value, it can be found that CPSO can reach the average fitness in a relatively short time. The number of iterations is 14 generations, and it takes 14.48 s. The traditional PSO algorithm requires 37 iterations and takes 12.84 s. The SHADE algorithm and wPSO require 26 and 37 iterations, consuming 14.21 s and 7.21 s. GA and ACA are obviously poor in fitness and higher than other algorithms. In summary, it can be proven that the CPSO algorithm is more stable and faster than other algorithms.

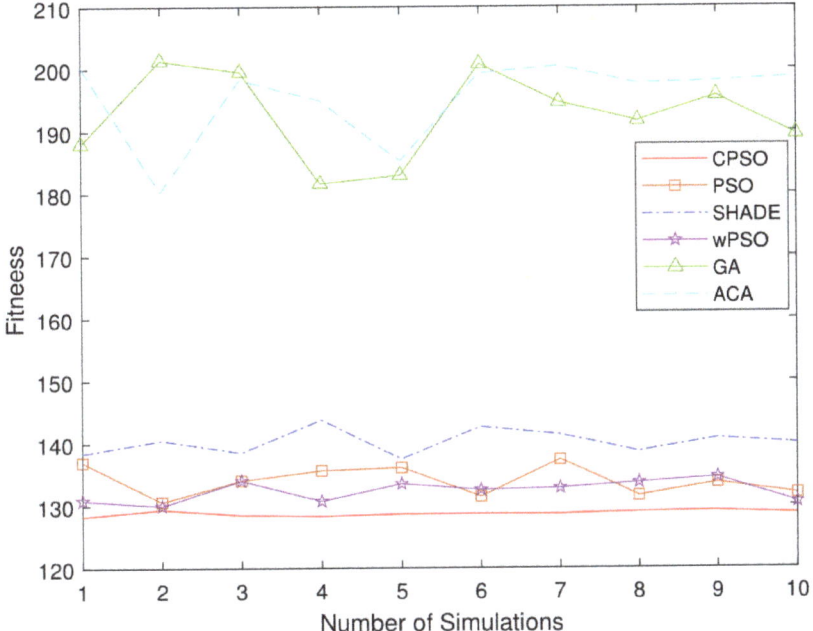

Figure 6. Ten simulation results.

Table 2. Simulation result statistics.

Algorithm	Average Fitness Value	Average Running Time/s	Fitness Value Variance	Average Number of Iterations to Reach Average Value	Time to Reach Average Fitness
CPSO	128.723	50.56	0.11	14	3.5
PSO	135.314	69.38	6.21	37	12.84
SHADE	140.237	90.45	4.011	60	27.14
wPSO	132.301	55.49	2.69	26	7.21
GA	192.585	85.28	44.941	34	14.5
ACA	195.355	275.38	47.337	39	53.7

5. Conclusions

Aiming at the shortcomings of the traditional PSO algorithm, with which it is easy to fall into local optimum, this paper proposes an improved PSO algorithm and applies it to 3D path planning. The improvement method is: introducing dynamic inertia weight, adding selection operation in genetic algorithm, adding crossover and mutation operation in SHADE algorithm, using mixed selection operations, introducing crossover and mutation operations, increasing the diversity of the population, retaining the global search ability of the particle swarm optimization algorithm, and enhancing the local search ability in the later stage of iteration. The CPSO algorithm, PSO algorithm, GA algorithm, and other algorithms are simulated and tested by MATLAB. The test results show that the improved PSO algorithm has better searching ability and stability, and performs significantly better in a complex environment than the comparison algorithm. Compared with PSO, GA, and other algorithms, the CPSO algorithm generates a shorter path length and smoother path, which improves the quality and efficiency of UAV routing. The CPSO average fitness value increased by about 5.12% compared with PSO, 49.61% compared with GA, 2.78% compared with wPSO, 8.94% compared with SHADE, and 51.76% compared with ACA.

This experiment verifies the feasibility of the algorithm through simulation, and assumes that the external environment has no influence. Future research will focus on the optimization of UAV three-dimensional path planning algorithms in real and dynamic environments and make them available for daily use.

Author Contributions: Writing—original draft, H.C.; Validation, X.Z.; Resources, H.L.; Writing—review and editing, L.D. All authors have read and agreed to the published version of the manuscript.

Funding: This research was funded by ZR2018QF005.

Conflicts of Interest: The authors declare no conflict of interest.

References

1. Nonami, K. Present state and future prospect of autonomous control technology for industrial drones. *IEEJ Trans. Electr. Electron Eng.* **2020**, *15*, 6–11. [CrossRef]
2. Giagkiozis, I.; Purshouse, R.C.; Fleming, P.J. An overview of population-based algorithms for multi-objective optimisation. *Int. J. Syst. Sci.* **2015**, *46*, 1572–1599. [CrossRef]
3. Deng, L.; Chen, H.; Liu, H.; Zhang, H.; Zhao, Y. Overview of UAV path planning algorithms. In Proceedings of the 2021 IEEE International Conference on Electronic Technology, Communication and Information (ICETCI), Changchun, China, 27–29 August 2021; pp. 520–523. [CrossRef]
4. Hart, P.E.; Nilsson, N.J.; Raphael, B. A Formal Basis for the Heuristic Determination of Minimum Cost Paths. *IEEE Trans. Syst. Sci. Cybern.* **1968**, *2*, 100–107. [CrossRef]
5. Dijkstra, E.W. A note on two problems in connexion with graphs. *Numer. Math.* **1951**, *1*, 269–271. [CrossRef]
6. Steinbrunn, M.; Moerkotte, G.; Kemper, A. Heuristic and randomized optimization for the join ordering problem. *VLDB J.* **1997**, *6*, 191–208. [CrossRef]
7. LaValle, S.M.; Kuffner, J.J. Randomized Kinodynamic Planning. *Int. J. Robot. Res.* **1999**, *15*, 378–400.
8. Wang, J.; Meng, Q.H. Optimal Path Planning Using Generalized Voronoi Graph and Multiple Potential Functions. *IEEE Trans. Ind. Electron.* **2020**, *67*, 10621–10630. [CrossRef]
9. Hua, X. Phase-out factor with particle swarm optimization. In Proceedings of the Second International Conference on Mechanic Automation and Control Engineering, Inner Mongolia, China, 17 July–15 July 2011.
10. Dorigo, M.; Gambardella, L.M. Ant colony system: A cooperative learning approach to the traveling salesman problem. *IEEE Trans. Evol. Comput.* **1997**, *1*, 53–66. [CrossRef]
11. Holland, J.H. Outline for a Logical Theory of Adaptive Systems. *J. ACM* **1962**, *9*, 297–314. [CrossRef]
12. Yang, L.; Qi, J.; Xiao, J.; Yong, X. A literature review of UAV 3D path planning. In Proceedings of the 11th World Congress on Intelligent Control and Automation, Shenyang, China, 29 June–4 July 2014.
13. Wang, Y.H.; Wang, S.M. UAV Path Planning Based on Improved Particle Swarm Optimization Algorithm. *Comput. Eng. Sci.* **2020**, *42*, 7.
14. Samuel, G.G.; Rajan, C.C.A. Hybrid Particle Swarm Optimization—Genetic algorithm and Particle Swarm Optimization—Evolutionary programming for long-term generation maintenance scheduling. In Proceedings of the 2013 International Conference on Renewable Energy and Sustainable Energy (ICRESE), Coimbatore, India, 5–6 December 2013.
15. Abhishek, B.; Ranjit, S.; Shankar, T.; Eappen, G.; Rajesh, A. Hybrid PSO-HSA and PSO-GA algorithm for 3D path planning in autonomous UAVs. *SN Appl. Sci.* **2020**, *2*, 1–16. [CrossRef]
16. Phung, M.; Ha, Q.P. Safety-enhanced UAV path planning with spherical vector-based particle swarm optimization. *Appl. Soft Comput.* **2021**, *107*, 107376. [CrossRef]
17. Qi, Z.; Shao, Z.; Ping, Y.S.; Hiot, L.M.; Leong, Y.K. An Improved Heuristic Algorithm for UAV Path Planning in 3D Environment. In Proceedings of the International Conference on Intelligent Human-Machine Systems & Cybernetics, Nanjing, China, 26–28 August 2010.

28. Kennedy, J.; Eberhart, R. Particle Swarm Optimization. In Proceedings of the Icnn95-International Conference on Neural Networks, Perth, WA, Australia, 27 November–1 December 1995.
29. Wang, X.; Meng, X. UAV Online Path Planning Based on Improved Genetic Algorithm. In Proceedings of the Chinese Control Conference (CCC), Guangzhou, China, 27–30 July 2019.
30. Wai, R.-J.; Prasetia, A.S. Adaptive Neural Network Control and Optimal Path Planning of UAV Surveillance System With Energy Consumption Prediction. *IEEE Access* **2019**, *7*, 126137–126153. [CrossRef]
31. Tanabe, R.; Fukunaga, A. Success-history based parameter adaptation for Differential Evolution. In Proceedings of the 2013 IEEE Congress on Evolutionary Computation (CEC), Cancun, Mexico, 20–23 June 2013.
32. Zhang, J.; Sanderson, A.C. JADE: Adaptive Differential Evolution with Optional External Archive. *IEEE Trans. Evol. Comput.* **2009**, *13*, 945–958. [CrossRef]
33. Tanabe, R.; Fukunaga, A.S. Improving the search performance of SHADE using linear population size reduction. In Proceedings of the Evolutionary Computation, Beijing, China, 6–11 July 2014.

Disclaimer/Publisher's Note: The statements, opinions and data contained in all publications are solely those of the individual author(s) and contributor(s) and not of MDPI and/or the editor(s). MDPI and/or the editor(s) disclaim responsibility for any injury to people or property resulting from any ideas, methods, instructions or products referred to in the content.

Article

A Novel Fixed-Time Convergence Guidance Law against Maneuvering Targets

Yaosong Long [1], Chao Ou [1,2], Chengjun Shan [1] and Zhongtao Cheng [1,*]

[1] School of Aerospace Engineering, Huazhong University of Science and Technology, Wuhan 430074, China
[2] Aerospace Technology Institute of China Aerodynamics Research and Development Center, Mianyang 621000, China
* Correspondence: ztcheng@hust.edu.cn

Abstract: In this paper, a new fixed-time convergence guidance law is proposed against maneuvering targets in the three-dimensional (3-D) engagement scenario. The fixed-time stability theory is used to zero the line-of-sight (LOS) angle rate, which will ensure the collision course and the impact of the target. It is proven that the convergence of the LOS angle rate can be achieved before the final impact time of the guidance process, regardless of the initial conditions. Furthermore, the convergence rate is merely related to control parameters. In theoretical analysis, the convergence rate and upper bound are compared with that of other laws to show the potential advantages of the proposed guidance law. Finally, simulations are carried out to illustrate the effectiveness and robustness of the proposed guidance law.

Keywords: three-dimensional engagement; maneuvering target; LOS angle rate; collision course; fixed-time convergence

MSC: 70E60; 93B12; 93C35; 93C85; 34A34

Citation: Long, Y.; Ou, C.; Shan, C.; Cheng, Z. A Novel Fixed-Time Convergence Guidance Law against Maneuvering Targets. *Mathematics* **2023**, *11*, 2090. https://doi.org/10.3390/math11092090

Academic Editors: Haizhao Liang, Jianying Wang and Chuang Liu

Received: 16 February 2023
Revised: 12 April 2023
Accepted: 21 April 2023
Published: 28 April 2023

Copyright: © 2023 by the authors. Licensee MDPI, Basel, Switzerland. This article is an open access article distributed under the terms and conditions of the Creative Commons Attribution (CC BY) license (https://creativecommons.org/licenses/by/4.0/).

1. Introduction

During a typical missile guidance process, the most important stage is the terminal guidance phase, which plays a decisive role in determining the missile's intercept performance [1]. In the terminal guidance phase, the target may be maneuvering. On the other hand, the time left for impacting the target is usually very short. Therefore, novel guidance laws with a fast convergence rate to ensure the impact and robustness to target maneuvers have great significance for the missile's performance.

In order to improve the missile's performance, many modern theories are utilized to design guidance laws. An effective way to improve the robustness of guidance laws is to apply the H_∞ control theory. In [2], the H_∞ guidance law was derived from solving the associated Hamilton–Jacobi function. In [3], based on the nonlinear robust H_∞ filtering method to estimate the LOS rate, a guidance law was proposed considering input saturation as well as system stability. Although strong robustness was obtained and exhibited, the H_∞ guidance laws cannot achieve finite-time convergence.

Another approach is to apply the Lyapunov asymptotic stability theory. In [4], a quadratic Lyapunov candidate function was proposed. By a particular selection of LOS angle function, the resulting guidance law can be free of singularities. In [5], a Lyapunov candidate function concerning the heading angle error was proposed, and the exact expression of the flight time was derived. The incomplete beta function was used, and the flight time can be adjusted by a single control parameter. As an improvement of the work in [5], another Lyapunov-based guidance strategy was proposed in [6] with impact angle constraints. In [7], the impact time constraint was considered, and the resulting guidance law can zero the heading error angle to ensure the collision course. In [8], a novel adaptive integrated guidance and control law was designed with a barrier Lyapunov function;

the resulting guidance law can handle input saturation and constraints of angles. This group of guidance laws was based on the Lyapunov asymptotic stable theory, and only guaranteed the convergence of the system as time approached infinitely. Moreover, the targets were assumed stationary in the design of the guidance law. Obviously, they are theoretically imperfect.

As an improvement of asymptotic convergence law, guidance laws that can ensure finite-time convergence were investigated. A guidance law based on finite-time convergence theory was proposed in [9], which was an early work considering both finite-time convergence and target maneuver together. The LOS rate converged to zero or a small neighborhood of zero in finite time under the proposed guidance law. Then, the work in [9] was improved by taking the autopilot dynamic into account [10,11]. In [12], based on sliding mode control theory, a finite-time convergence guidance law with impact angle constraint was proposed, and the guidance command was generated to enforce terminal sliding mode on the designed switching surface from nonlinear engagement dynamics. As an improvement of [12], the work in [13] was based on the output feedback continuous terminal sliding mode guidance. The resulting guidance law can achieve not only finite-time convergence but also ensure continuity of control action. Compared with guidance laws based on stable asymptotic theory, this group of guidance laws can achieve finite-time convergence. However, the convergence upper bound is relative to initial states.

Recently, the design of guidance law also applied the fixed-time stability theory, which was an improvement of finite-time stability theory since it can provide a settling upper bound irrelevant to initial conditions. Since this theory was first presented in [14] in 2012, few works have utilized this theory for guidance law design. The earliest work in this direction was found in [15], where a planar adaptive fixed-time guidance law was presented; the resulting guidance law can stabilize the guidance system with a bounded settling time without dependence on the initial conditions. Then, the work in [15] was improved by considering time constraints with input delays [16]. The fixed-time stability was further applied to the 3-D engagement scenario [17,18]. The work in [17] utilized the fixed-time stability theory to achieve a fast consensus protocol. Then, it was improved in [18] by considering the impact angle constraint. Despite the settling time being irrelative to initial conditions, the guarantee of the settling time before the final impact time is not discussed by the above-mentioned guidance laws. The fixed-time consensus tracking algorithms of second-order MASs via event-triggered control are presented in [19]; for the fixed-time consensus result, the consensus can be reached in a settling time with any initial condition, and it is revealed that the ratio of each pair of states is constant resulting in shorter output trajectories [20,21] investigate the fixed-time synchronization problem for the coupled neural networks, respectively. Recently, [22] has given the concept of practically fixed-time stability for the first time. The finite/fixed-time stabilization and tracking control problems are simultaneously concerned in [23–25].

Inspired by the above observation, this paper proposes the fixed-time convergence guidance law against maneuvering targets in 3-D engagement scenarios. It is proven that the convergence of the LOS angle rate to zero can be completed before the final impact time, regardless of the initial conditions. To the best of the authors' knowledge, guidance laws consider the following three problems simultaneously, i.e., fixed-time convergence, 3-D engagement against maneuvering target, and the guarantee of the settling time before the final impact time, which are rare in the literature.

The main contribution of this work can be stated as follows:

(1) A fixed-time convergence guidance law for 3-D engagement scenarios is proposed against the maneuvering target. The novel guidance law can ensure fixed-time convergence and fixed-time stability without initial condition constraints.

(2) The settling time of the LOS rate is proven to be surely shorter than the minimum final impact time by the proposed fixed-time guidance law. It can ensure the success of the missile in hitting the maneuvering target.

(3) The convergence rate is proven merely related to control parameters, a suitable selection of which can ensure the convergence rate without violating acceleration constraints.

The following of this paper is structured as follows. The homing guidance model for the 3-D engagement scenario and the main objective of the guidance law are introduced in Section 2, respectively. The Fixed-time convergent guidance law design and the analysis of its property are offered in Section 3. Simulations are carried out in Section 4 to show the effectiveness of the proposed guidance laws. Finally, the conclusion of the work is proposed in Section 5.

2. Problem Formulation

In this section, first, the dynamic model describing the motion of the aerospace vehicle is offered. Then, the main objective of the guidance law is introduced.

Homing Guidance Model

The guidance geometry in 3-D space is constructed in Figure 1, where $MXYZ$ is the inertial reference coordinate and $Mxyz$ is the LOS coordinate. M represents the missile and T denotes the target. r denotes the relative range between the missile and the target. ϕ and θ are the azimuth and elevation LOS angle, respectively. The angles in Figure 1 are measured positively in the counterclockwise direction.

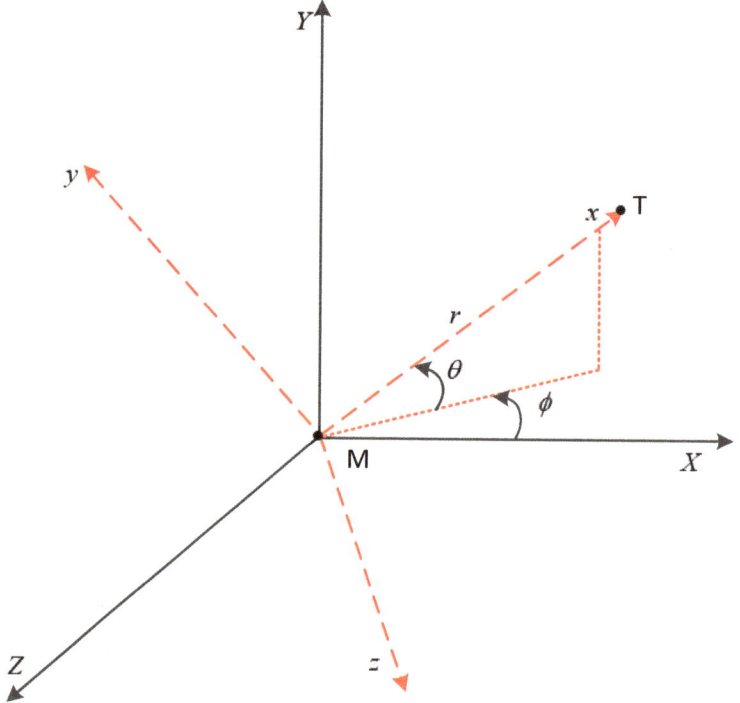

Figure 1. Three-dimensional missile-target interception geometry.

Define $a_T = (a_{Tr}, a_{T\theta}, a_{T\phi})$ and $a_M = (a_{Mr}, a_{M\theta}, a_{M\phi})$ as the accelerations measured in the LOS frame for the target and the missile, respectively. According to the virtue of kinematics, the relative velocity V between the missile and the target can be expressed as

$$V(R, \theta, \phi) = \begin{bmatrix} \dot{R} \\ R\dot{\theta} \\ -R\dot{\phi}\cos\theta \end{bmatrix} \tag{1}$$

According to the derivative rule of vector derivatives, we have

$$\frac{dV}{dt} = \omega \times V + \frac{\partial V}{\partial t} = a_T - a_M \tag{2}$$

where dV/dt and $\partial V/\partial t$ refers to the derivate of V in $MXYZ$ and $Mxyz$, respectively, and ω represents the rotation speed for $Mxyz$ relative to the inertial coordinate system $MXYZ$. It can be acquired from Figure 1 that

$$\omega \times = \begin{bmatrix} 0 & -\dot{\theta} & \dot{\phi}\cos\theta \\ \dot{\theta} & 0 & -\dot{\phi}\sin\theta \\ -\dot{\phi}\cos\theta & \dot{\phi}\sin\theta & 0 \end{bmatrix} \tag{3}$$

By substituting Equations (1) and (3) into Equation (2), we can obtain

$$\ddot{r} - r\dot{\phi}^2 - r\dot{\theta}^2\cos^2\phi = a_{Tr} - a_{Mr} \tag{4}$$

$$r\ddot{\theta}\cos\phi + 2\dot{r}\dot{\theta}\cos\phi - 2r\dot{\phi}\dot{\theta}\sin\phi = a_{T\theta} - a_{M\theta} \tag{5}$$

$$r\ddot{\phi} + 2\dot{r}\dot{\phi} + r\dot{\theta}^2\sin\phi\cos\phi = a_{T\phi} - a_{M\phi} \tag{6}$$

It should be noted that the collision course can be achieved with the LOS angle rate $\dot{\theta}$ and $\dot{\phi}$ converging to zero before hitting the target. Thus, Equations (5) and (6) are considered in the design of the guidance law.

Define $x_1 = \dot{\theta}$, $x_2 = \dot{\phi}$, $u_1 = a_{M\theta}$, $u_2 = a_{M\phi}$, and the coupling state equations of LOS angles can be acquired as:

$$\dot{x}_1 = -\frac{2\dot{r}}{r}x_1 + 2x_1 x_2 \tan\phi - \frac{u_1}{r\cos\phi} + \frac{a_{T\theta}}{r\cos\phi} \tag{7}$$

$$\dot{x}_2 = -\frac{2\dot{r}}{r}x_2 - x_1^2 \sin\phi\cos\phi - \frac{u_2}{r} + \frac{a_{T\phi}}{r} \tag{8}$$

It can be concluded from Equation (7) that there exists cross-coupling between $\dot{\theta}$ and $\dot{\phi}$. By virtue of the analysis in [1], x_1 and x_2 are small variables during the time horizon of the impact process. This gives $\cos\phi \approx 1$. Moreover, the third order of the small variables can be neglected. Hence, Equation (7) can be rewritten as

$$\dot{x}_1 = -\frac{2\dot{r}}{r}x_1 - \frac{u_1}{r} + \frac{a_{T\theta}}{r} \tag{9}$$

$$\dot{x}_2 = -\frac{2\dot{r}}{r}x_2 - \frac{u_2}{r} + \frac{a_{T\phi}}{r} \tag{10}$$

The primary objective of the guidance law is to hit the target, which can be achieved with the convergence of the LOS angle rates converging to zero in both planes. Therefore, the objective is to design the guidance law that can zero the LOS angle rates before hitting the target.

3. Guidance Law Design

Considering the 3-D LOS angle motions are decoupled into two (2-D) LOS angular motions in the previous analysis, in this section, the planar fixed-time convergence guidance law is presented first. Then, the planar guidance law is further applied to the 3-D scenario Although the decoupled model is utilized in the design of the 3-D guidance law, the proof for the convergence of LOS angle rates conducts on the cross-coupling model directly.

3.1. The Planar Guidance Law Design

In this subsection, the planar guidance law that can zero the LOS rate before hitting the target is proposed, and the decouple planar LOS motion of Equation (9) is considered in the design of the guidance law. Before deriving the guidance law, it is obliged to introduce some basic lemma of fixed-time stability theory.

Before deriving the guidance law, it is obliged to introduce some basic concepts of fixed-time stability theory [14].

Definition: The following nonlinear system is considered:

$$\dot{x}(t) = f(t, x(t)), \quad x(0) = x_0 \qquad (11)$$

where the state and the upper semi-continuous mapping are denoted by $x(t) \in R^l$ and $f : R^+ \times R^n \to R^n$, respectively. The state is fixed-time stability if it is globally finite-time stable. Meanwhile, the function of the settling time $T(x_0)$ is restricted by a real positive number T_{max}, i.e., $T(x_0) \leq T_{max}, \forall x_0 \in R^l$. The definition can be stated mathematically as

$$\begin{cases} \lim_{t \to T(x_0)} x(t, x_0) = 0. & t \in [t_0, T(x_0)) \\ x(t, x_0) = 0. & t \geq T(x_0), T(x_0) < T_{max} \end{cases} \qquad (12)$$

Denote $D^*\varphi(t)$ as the upper right-hand derivative of a function $\varphi(t)$, $D^*\varphi(t) = \lim_{h \to +0}(\varphi(t+h) - \varphi(t))/h$. The fixed-time stability under the Lyapunov criterion is presented in Lemma 1.

Lemma 1. *Suppose a continuous positive definite and radially unbounded function as $V(x) : R^n \to R^+ \cup \{0\}$, such that:*

$$D^*V(x(t)) \leq -mV^p(x(t)) - nV^q(x(t)) \qquad (13)$$

for $m, n > 0, p = 1 - \frac{1}{2\gamma}, q = 1 + \frac{1}{2\gamma}, \gamma > 1$, then the origin is fixed-time stable for the system $V(x)$, and the settling time is given by:

$$T(x_0) \leq T_{max} := \frac{\pi\gamma}{\sqrt{mn}} \qquad (14)$$

Assume the deviations from the collision course for both the missile and target are small, then the relative velocity can be approximated as:

$$\dot{r} = -c, \quad c = \text{const.} > 0 \qquad (15)$$

This assumption is reasonable since it can be conducted by a well-midcourse guidance process. Then, the instant range at time t can be acquired as

$$r(t) = r_0 - ct \qquad (16)$$

Theorem 1. *If the guidance command u_1 can make the LOS angle rate x_1 satisfying:*

$$x_1[\dot{x}_1 + \frac{m|x_1|^{1-\frac{1}{\gamma}}\mathrm{sgn}(x_1) + n|x_1|^{1+\frac{1}{\gamma}}\mathrm{sgn}(x_1)}{2r(t)}] \leq 0 \quad (17)$$

where $m = $ const. > 0, $n = $ const. > 0, $\gamma = $ const. > 1, and

$$\mathrm{sgn}(x) = \begin{cases} 1, & x \geq 0 \\ -1, & x \leq 0 \end{cases} \quad (18)$$

Then, $\dot{x}_1$ will converge to zero before hitting the target.

Proof. The following continuously differential candidate function is considered:

$$W_1 = x_1^2 \quad (19)$$

The derivative of Equation (19) to time is

$$\dot{W}_1 = 2x_1\dot{x}_1 \quad (20)$$

Substituting Equation (17) into Equation (20) yields

$$\dot{W}_1 \leq -\frac{m}{r}W_1^{1-\frac{1}{2\gamma}} - \frac{n}{r}W_1^{1+\frac{1}{2\gamma}} \quad (21)$$

According to Lemma 1, W_1 will converge to zero in fixed-time. Define the settling time for W_1 as T_1, then we have

$$\lim_{t \to T_1} W_1 = 0 \quad (22)$$

Since $W_1 = 0$ in Equation (21) is a trivial case, assuming $W_1 \neq 0$ yields

$$\frac{dW_1}{dt} \leq -\frac{mW_1^{1-\frac{1}{2\gamma}} + nW_1^{1+\frac{1}{2\gamma}}}{r} \quad (23)$$

Substituting Equation (13) into (23) yields:

$$\frac{dW_1}{mW_1^{1-\frac{1}{2\gamma}} + nW_1^{1+\frac{1}{2\gamma}}} \leq -\frac{dt}{r_0 - ct} \quad (24)$$

Integrating the right side of Equation (24) from 0 to T_1, and the corresponding integral interval for the left side is $[W_1(0), W_1(T_1)]$. One can obtain

$$\int_{W_1(0)}^{W_1(T_1)} \frac{1}{mW_1^{1-\frac{1}{2\gamma}} + nW_1^{1+\frac{1}{2\gamma}}} dW_1 \leq \frac{1}{c}\ln(1 - \frac{cT_1}{r_0}) \quad (25)$$

Define

$$\varphi = -\int_{W_1(0)}^{W_1(T_1)} \frac{c}{mW_1^{1-\frac{1}{2\gamma}} + nW_1^{1+\frac{1}{2\gamma}}} dW_1 \quad (26)$$

Substituting Equation (26) into (25) yields

$$T_1 \leq (1 - \frac{1}{e^\varphi})\frac{r_0}{c} \quad (27)$$

Define t_f as the final time of the engagement. According to Equation (16), one can obtain

$$t_f = \frac{r_0}{c} \quad (28)$$

Then, Equation (27) can be rewritten as

$$T_1 \leq (1 - \frac{1}{e^\varphi})t_f \tag{29}$$

where φ merely relates to the initial LOS rate. Substituting Equation (22) into (26), one can further obtain

$$\varphi = \int_{W_1(T_1)}^{W_1(0)} \frac{c}{mW_1^{1-\frac{1}{2\gamma}} + nW_1^{1+\frac{1}{2\gamma}}} dW_1 \tag{30}$$

Thus, we can obtain $\varphi > 0$. Combining Equations (29) and (30) yields $T_1 < t_f$, which implies that the convergence time for the LOS rate is always less than t_f regardless of the initial conditions. Hence, the proof of Theorem 1 is completed. □

Substituting Equation (9) into (17) yields

$$x_1[-\frac{2\dot{r}}{r}x_1 + \frac{1}{r}u_1 - \frac{1}{r}a_{T\theta} + \frac{m}{2r}|x_1|^{1-\frac{1}{\gamma}}\text{sgn}(x_1) + \frac{n}{2r}|x_1|^{1+\frac{1}{\gamma}}\text{sgn}(x_1)] \leq 0 \tag{31}$$

Then, the guidance command is chosen a

$$u_1 = \begin{cases} -N\dot{r}x_1 + r(\frac{m}{2}|x_1|^{1-\frac{1}{\gamma}} + \frac{n}{2}|x_1|^{1+\frac{1}{\gamma}})\text{sgn}(x_1) + a_{T\theta}, & r \geq 1 \\ -N\dot{r}x_1 + (\frac{m}{2}|x_1|^{1-\frac{1}{\gamma}} + \frac{n}{2}|x_1|^{1+\frac{1}{\gamma}})\text{sgn}(x_1) + a_{T\theta}, & 0 \leq r < 1 \end{cases} \tag{32}$$

where $N = \text{const.} > 2$, and the additional term r acts as the adaptive term to speed up the convergence process before hitting the target.

Theorem 2. *The guidance command in Equation (32) can zero the LOS angle rate before hitting the target.*

Proof. Substitute Equation (32) into (9), we have

$$\dot{x}_1 = \begin{cases} \frac{(N-2)\dot{r}x_1}{r} - (\frac{m}{2}|x_1|^{1-\frac{1}{\gamma}} + \frac{n}{2}|x_1|^{1+\frac{1}{\gamma}})\text{sgn}(x_1), & r \geq 1 \\ \frac{(N-2)\dot{r}x_1}{r} - (\frac{m}{2r}|x_1|^{1-\frac{1}{\gamma}} + \frac{n}{2r}|x_1|^{1+\frac{1}{\gamma}})\text{sgn}(x_1), & 0 \leq r \leq 1 \end{cases} \tag{33}$$

By substituting Equation (33) into (17) yields

$$\begin{cases} \frac{(N-2)\dot{r}x_1^2}{r} + \frac{m}{2}|x_1|^{2-\frac{1}{\gamma}}(\frac{1}{r}-1) + \frac{n}{2}|x_1|^{2+\frac{1}{\gamma}}(\frac{1}{r}-1) \leq 0, & r \geq 1 \\ \frac{(N-2)\dot{r}x_1^2}{r} \leq 0, & 0 \leq r < 1 \end{cases} \tag{34}$$

According to Theorem 1, the proposed guidance command in Equation (32) can lead to fixed-time convergence for the LOS angle rate, and the convergence rate increases as the value of m and n increases, or as the value of γ decreases. Hence, the proof of Theorem 2 is completed. □

3.2. Guidance Law Design in 3-D Engagement Scenario

According to the planar guidance law designed in the previous section, the fixed-time convergence guidance command for the 3-D engagement scenario can be designed as

$$\begin{cases} u_1 = \begin{cases} -N\dot{r}x_1 + a_{T\theta} + r(\frac{m}{2}|x_1|^{1-\frac{1}{\gamma}} + \frac{n}{2}|x_1|^{1+\frac{1}{\gamma}})\text{sgn}(x_1), & r \geq 1 \\ -N\dot{r}x_1 + a_{T\theta} + (\frac{m}{2}|x_1|^{1-\frac{1}{\gamma}} + \frac{n}{2}|x_1|^{1+\frac{1}{\gamma}})\text{sgn}(x_1), & 0 \leq r \leq 1 \end{cases} \\ u_2 = \begin{cases} -N\dot{r}x_2 + a_{T\phi} + r(\frac{m}{2}|x_2|^{1-\frac{1}{\gamma}} + \frac{n}{2}|x_2|^{1+\frac{1}{\gamma}})\text{sgn}(x_2), & r \geq 1 \\ -N\dot{r}x_1 + a_{T\phi} + (\frac{m}{2}|x_1|^{1-\frac{1}{\gamma}} + \frac{n}{2}|x_1|^{1+\frac{1}{\gamma}})\text{sgn}(x_1), & 0 \leq r \leq 1 \end{cases} \end{cases} \tag{35}$$

Theorem 3. *The guidance commands in Equation (35) can achieve fixed-time convergence for the LOS angle rates in Equations (7) and (8) before hitting the target.*

Proof. The proof of Theorem 3 is divided into two parts. First, the effectiveness of the proposed guidance command under the condition $0 < r \leq 1$ is proven.

Substituting Equation (35) into (7) and (8) yields

$$\begin{aligned}\dot{x}_1 &= \frac{(N/\cos\phi - 2)\dot{r}}{r}x_1 + 2x_1 x_2 \tan\phi - \text{sgn}(x_1)(m|x_1|^{1-\frac{1}{\gamma}} + n|x_1|^{1+\frac{1}{\gamma}})/2r\cos\phi \\ \dot{x}_2 &= \frac{(N-2)\dot{r}}{r}x_2 - x_1^2 \sin\phi\cos\phi - \text{sgn}(x_2)(m|x_2|^{1-\frac{1}{\gamma}} + n|x_2|^{1+\frac{1}{\gamma}})/2r\cos\phi\end{aligned} \quad (36)$$

The following continuously differential candidate function is considered:

$$W_2 = x_1^2 \cos^2\phi + x_2^2 \quad (37)$$

The derivative of Equation (37) with respect to time is

$$\dot{W}_2 = 2x_1 \dot{x}_1 \cos\phi + 2x_2 \dot{x}_2 - 2x_1^2 x_2 \sin\phi \cos\phi \quad (38)$$

Substituting Equation (36) into (38) yields

$$\begin{aligned}\dot{W}_2 &= \frac{2(N/\cos\phi-2)\dot{r}x_1^2 \cos^2\phi}{r} - \text{sgn}(x_1)(m|x_1|^{2-\frac{1}{\gamma}} + n|x_1|^{2+\frac{1}{\gamma}})\frac{\cos\phi}{r} \\ &+ \frac{2(N-2)\dot{r}}{r}x_2^2 - \text{sgn}(x_2)(\frac{m}{r}|x_2|^{2-\frac{1}{\gamma}} + \frac{n}{r}|x_2|^{2+\frac{1}{\gamma}})\end{aligned} \quad (39)$$

Since

$$\frac{2(N/\cos\phi - 2)\dot{r}x_1^2 \cos^2\phi}{r} \leq 0 \quad (40)$$

Then, we can obtain

$$\dot{W}_2 \leq -(mx_1^{2-\frac{1}{\gamma}} + nx_1^{2+\frac{1}{\gamma}})\frac{\cos\phi}{r} - (mx_2^{2-\frac{1}{\gamma}} + nx_2^{2+\frac{1}{\gamma}})\frac{1}{r} \quad (41)$$

By choosing an appropriate inertial reference coordinate system, we can ensure that $-0.5\pi < \cos\phi < 0.5\pi$. Thus, $0 < \cos\phi < 1$. Then, Equation (41) can be rewritten as:

$$\begin{aligned}\dot{W}_2 &\leq -[\frac{m}{r}(x_1 \cos\phi)^{2-\frac{1}{\gamma}} + \frac{m}{r}x_2^{2-\frac{1}{\gamma}}] - [\frac{n}{r}(x_2 \cos\phi)^{2+\frac{1}{\gamma}} + \frac{n}{r}x_2^{2+\frac{1}{\gamma}}] \\ &\leq -\frac{m}{r}(x_1^2 \cos^2\phi + x_2^2)^{1-\frac{1}{2\gamma}} - \frac{n}{r}(x_1^2 \cos^2\phi + x_2^2)^{1+\frac{1}{2\gamma}}\end{aligned} \quad (42)$$

As we define in Theorem 1 that $\gamma = \text{const.} > 1$, we can further obtain

$$\dot{W}_2 \leq -\frac{m}{r}(x_1^2 \cos^2\phi + x_2^2)^{1-\frac{1}{2\gamma}} - \frac{n}{r}(x_1^2 \cos^2\phi + x_2^2)^{1+\frac{1}{2\gamma}} \quad (43)$$

which can be written in an alternative form as

$$\dot{W}_2 \leq -\frac{m}{r}V^{1-\frac{1}{2\gamma}} - \frac{n}{r}V^{1+\frac{1}{2\gamma}} \quad (44)$$

The proof for the proposed guidance command under the condition $r \geq 1$ is similar; thus, it is omitted here. According to Theorem 1, the proposed guidance commands in Equation (35) can achieve fixed-time convergence for the LOS angle rate in Equations (7) and (8). Define T_2 as the convergence time in a 3-D scenario, the upper bound of the convergence time is given by

$$T_2 \leq T_{m_2} = (1 - \frac{1}{e^{\varphi_m}})t_f \quad (45)$$

where T_{m2} is the upper bound for the settling time for the 3-D guidance scenario. It is obvious that T_{m2} is independent of the initial states. □

3.3. Discussion of the Potential Advantage of the Proposed Guidance Law

To facilitate the comparison between different guidance laws, the variable that needs to be restrained to zero during the guidance process is defined as ε. Some results on the design of guidance laws adopt the Lyapunov asymptotic stability theory as [6], the dynamic of ε is

$$\dot{\varepsilon} = \frac{kV \sin \varepsilon_0}{r_0^k} r^{k-1}, \ k > 1 \tag{46}$$

Theoretically, the Lyapunov asymptotic stability theory only guarantees the convergence of ε when the time approaches infinity, and Equation (46) implies that $\dot{\varepsilon}(t) = 0$ when and only when $r = 0$, which means the convergence process of ε completes exactly at the instant of hitting the target. Some ideal assumptions are made during the design process of the guidance law. On the other hand, uncertainties and disturbances exist in practical applications. Hence, the error dynamic in Equation (46) may fail to converge to zero at the terminal instant in practical applications. Compared with the guidance law in [6], the proposed guidance law can ensure the convergence of ε before hitting the target, which makes it more robust to uncertainties and disturbances.

Some other results are based on the finite-time stability theory as [11], the convergence of ε can be completed in finite time, and the settling time satisfies

$$T < \frac{|\varepsilon_0|^{1-\eta} r_0}{\beta(1-\eta)} \tag{47}$$

where $\beta = const. > 0$ and $0 \leq \eta = const. < 1$.

Compared with guidance laws based on the Lyapunov asymptotic stability theory, this group of guidance laws can achieve finite-time convergence of ε. However, the convergence upper bound in Equation (47) depends on initial states, and only a proper selection of the control parameter can ensure ε converged to zero before the final interception time. By contrast, the proposed guidance law can ensure convergence before hitting the target regardless of the initial conditions.

4. Simulations

In this section, numerical simulations are carried out to show the effectiveness of the proposed guidance laws. All the simulations are conducted on the Matlab platform via C++ programming. The simulation step is 0.01 s. All the simulations are terminated when the sign of the relative velocity becomes positive, or the relative range is less than 0.01 m.

4.1. Comparison Simulations

In this case, the comparison simulation is considered to show the effectiveness of the proposed guidance law. Detailed simulation parameters are tabulated in Table 1.

Table 1. Initial states for the missile.

Parameter	Value
Initial missile position	(0, 0) m
Missile speed	300 m/s
Missile's initial heading angle	80°

The performance of the proposed guidance law is compared with the finite-time convergence guidance law (FTCG) proposed in [9]. The guidance command for FTCG is given by

$$a_M^{FTCG}(t) = -C R \dot{\lambda} + \beta \left| \dot{\lambda} \right|^\eta \text{sgn}(\dot{\lambda}) \tag{48}$$

where $C = const. > 2$, $\beta = const. > 0$, $0 \leq \eta = const. < 1$.

Two different control parameters are selected for the comparison law. Detailed control parameters, in this case, are summarized in Table 2.

Table 2. Control parameters for guidance laws.

Guidance Law	Value
FTCG1	$\beta = 20, \eta = 0.1$
FTCG2	$\beta = 20, \eta = 1$
Proposed	$m = n = 0.2, \gamma = 5$

Simulation results for both guidance laws are shown in Figure 2. Dot lines represent the results of the proposed guidance law. Dash lines and solid lines represent the results for FTCG under two different control parameters. Figure 2a shows the elevation acceleration, and Figure 2b represents the profile of the elevation LOS angle rate. Figure 2c shows the azimuth acceleration, and Figure 2d represents the profile of the azimuth LOS angle rate.

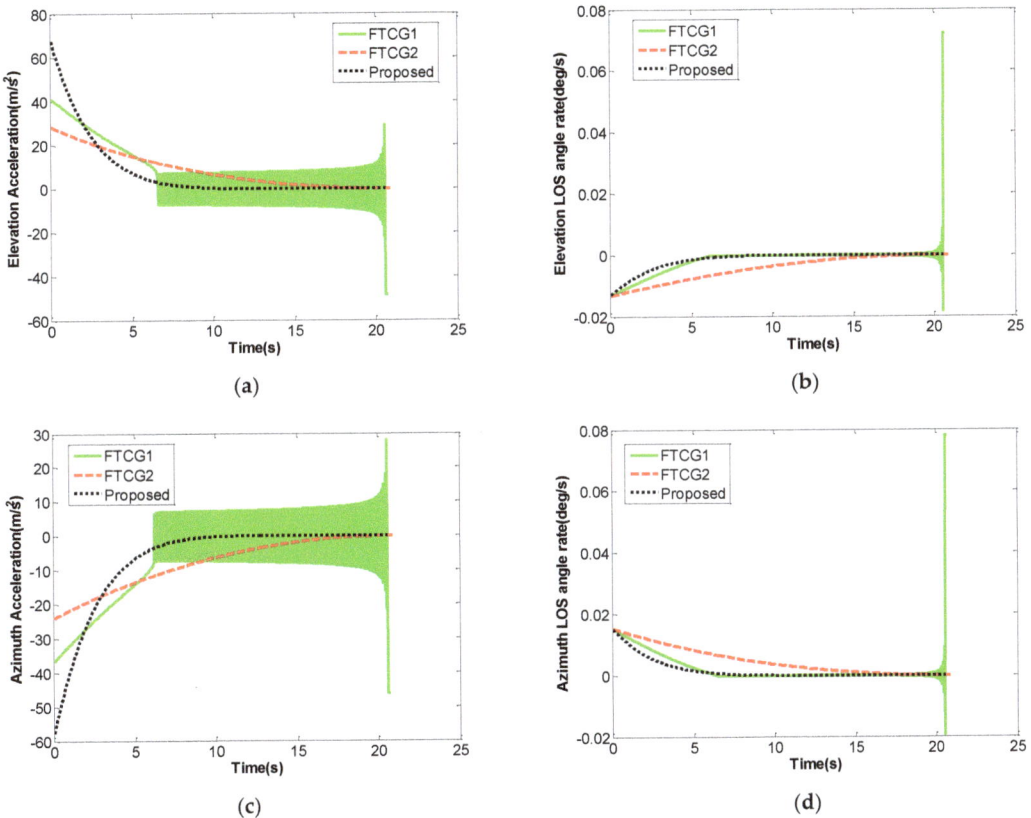

Figure 2. Comparison results. (a) Elevation acceleration. (b) Elevation LOS angle rate. (c) Azimuth acceleration. (d) Azimuth LOS angle rate.

Although each guidance law can impact the target successfully, the acceleration variation and the convergence of the LOS angle rate are significantly different. The acceleration for the proposed guidance law converges to zero in fixed time and remains there afterward, while the comparison law with the first group of control parameters will fluctuate around zero until the instant of impact, as demonstrated by FTCG1. There would be no chattering for the proposed guidance law under any allowable control parameters. As a result, the proposed guidance law can achieve higher accuracy than the comparison law. However, the comparison law can avoid chattering by proper selection of control parameters, as FTCG2 does. However, the LOS angle rate only converges to zero at the end of the impact, failing to exhibit the characteristic of finite-time convergence. Hence, the proposed guidance law has better performance than the comparison law.

4.2. Simulations with Autopilot Dynamics

As shown in the previous simulation case, the initial acceleration for the missile under the proposed law is very large. However, acceleration usually grows from scratch in practice. Furthermore, the autopilot delays are uncompensated during the design of the guidance law. Hence, it is necessary to investigate the performance of the proposed guidance law under the effect of autopilot dynamics.

Some existing methods compensate for the autopilot dynamics by computing the control parameter at each time step in a feedback manner, which makes the guidance law more complicated. Since robustness is a generic characteristic of the proposed guidance law, the control parameters do not need to be calculated in a feedback-step manner. To show the robustness of the proposed guidance laws, a first-order autopilot dynamic is considered in this simulation, which can be expressed as

$$\frac{a_{qa}}{a_q} = \frac{1}{1+\tau s} \qquad (49)$$

where a_q is the ideal acceleration, a_{qa} is the actual acceleration. The time constant τ considered for the autopilot dynamic, in this case, is $0.5s$. Initial conditions and control parameters are the same as in the previous section.

Simulation results are shown in Figure 3. Ideal and actual accelerations are plotted with different types of lines in Figure 3a,c. It is obvious that there exists a tracking error between the ideal and actual accelerations under the effect of autopilot delays. However, this error can be eliminated in a fixed time without extra effort under the proposed guidance law. Simulation results with actual command are plotted in blue solid line in Figure 3b,d, which converge to zero before the final time and ensure the successful impact of the target. Even though the proposed guidance law is derived from a lag-free system, the guidance law can provide high accuracy in a realistic missile system with autopilot lag.

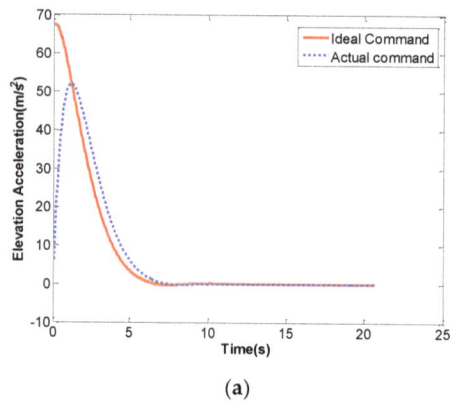

(a)

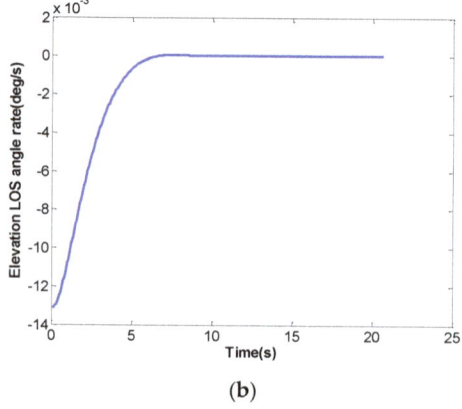

(b)

Figure 3. Cont.

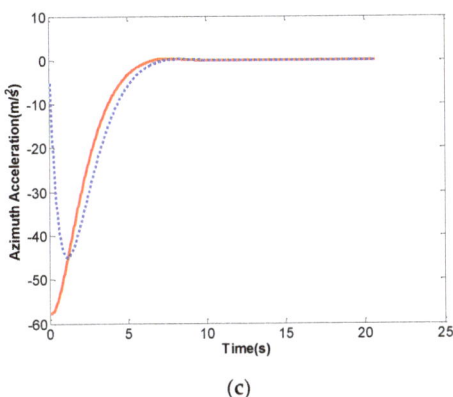
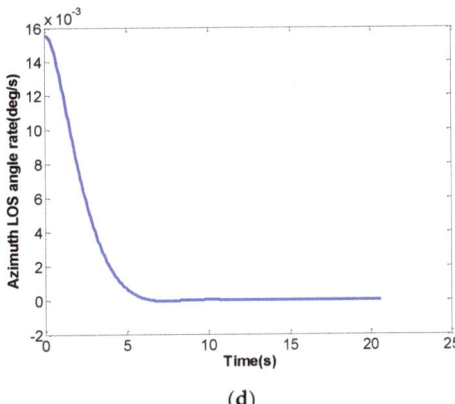

(c) (d)

Figure 3. Simulation with autopilot dynamics. (**a**) Elevation acceleration. (**b**) Elevation LOS angle rate. (**c**) Azimuth acceleration. (**d**) Azimuth LOS angle rate.

4.3. Simulations with Different m and n

In this case, the performance of the proposed guidance law in a 3-D Scenario is studied under different control parameters, which are $m = n = 0.1, m = n = 0.2, m = n = 0.4$.

Simulation results are shown in Figures 4 and 5. Solid lines, dash lines, and dot lines represent the results for three different control parameters. Figure 4a shows the elevation acceleration, and Figure 4b represents the profile of the elevation LOS angle rate. Figure 5a shows the azimuth acceleration, and Figure 5b represents the profile of the azimuth LOS angle rate.

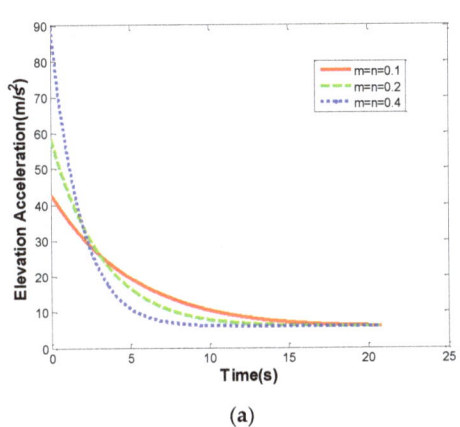
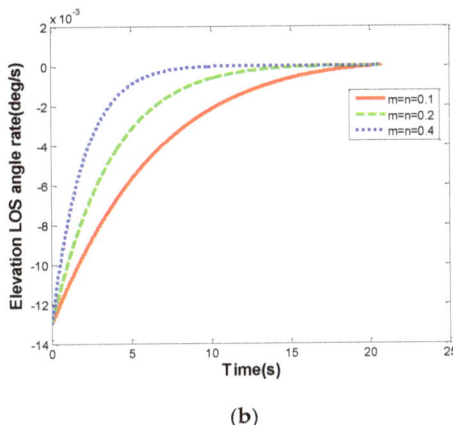

(a) (b)

Figure 4. Simulation results in the elevation plane with different m and n. (**a**) Elevation acceleration. (**b**) Elevation LOS angle rate.

For all the various values of control parameters, the LOS angle rate can converge to zero in fix time, as shown in Figures 4b and 5b. The collision course can be achieved, and the impact of the target can be ensured. Moreover, the miss distance for the missile can be less than 0.1 m. It also can be concluded from Figures 4b and 5b that the convergence rate for the LOS angle rate increases as the value of the control parameter increases. This is in line with Theorem 3. However, a higher convergence rate requires larger guidance commands at the beginning of the guidance process, as is demonstrated in Figures 4a and 5a.

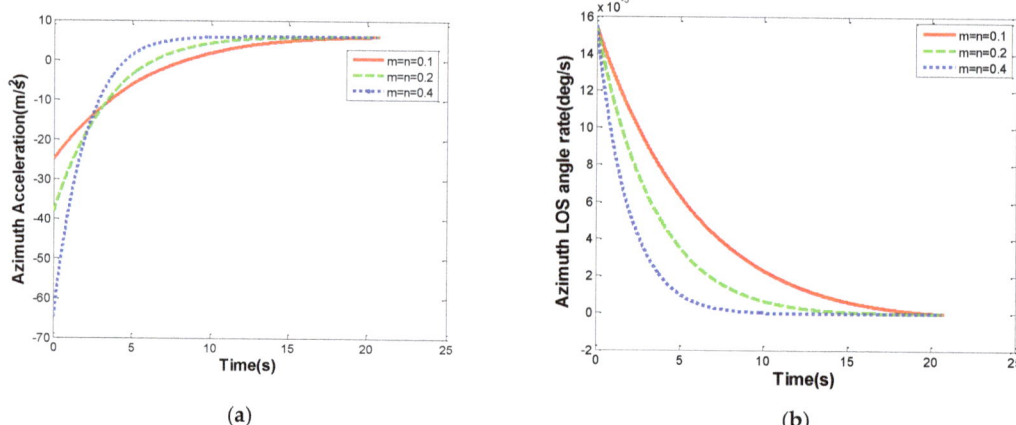

Figure 5. Simulation results in the azimuth plane with different m and n. (**a**) Azimuth acceleration. (**b**) Azimuth LOS angle rate.

4.4. Simulations with Different γ

In this case, the initial conditions for the missile and the initial coordinates for the target are the same as in Section 4.1. The speed for the target is $V_T = 200$ m/s, and the acceleration is $a_T = 6$ m/s.

Simulation results are shown in Figures 6 and 7. Solid lines, dash lines, and dot lines represent the results for three different control parameters. Figure 6a shows the elevation acceleration, and Figure 6b represents the profile of the elevation LOS angle rate. Figure 7a shows the azimuth acceleration, and Figure 7b represents the profile of the azimuth LOS angle rate. The collision course is achieved with the LOS angle rate converging to zero. It is clear from Figures 6b and 7b that the convergence of the LOS angle rate achieves in fixed time for all the control parameters. It also can be concluded from Figures 6b and 7b that the convergence rate for the LOS angle rate increases as the value of the control parameter decreases. This is in line with Theorem 3. Moreover, a higher convergence rate requires larger guidance commands at the beginning of the guidance process, as is demonstrated in Figures 6a and 7a.

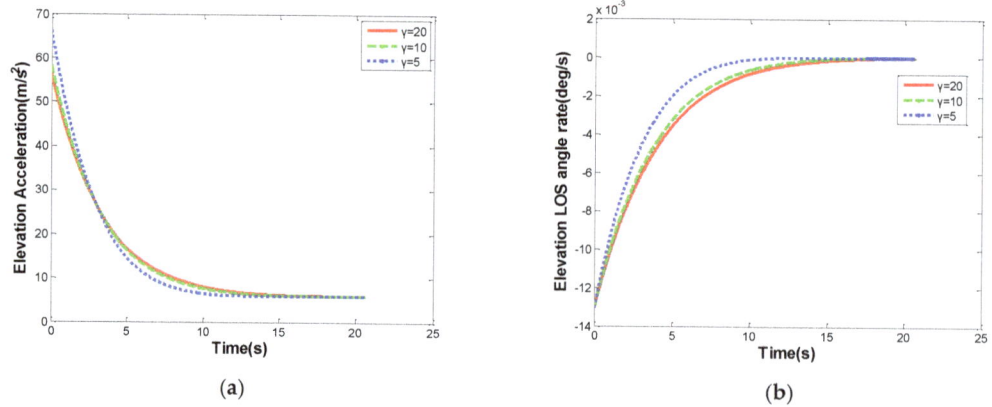

Figure 6. Simulation results in the elevation plane with different γ. (**a**) Elevation acceleration. (**b**) Elevation LOS angle rate.

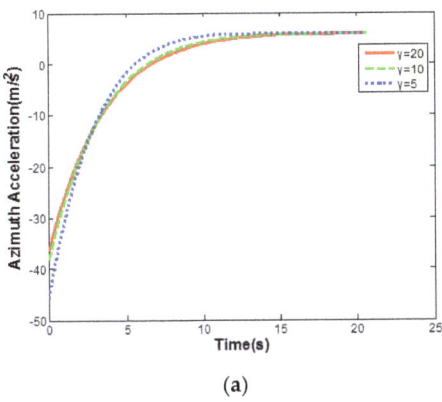
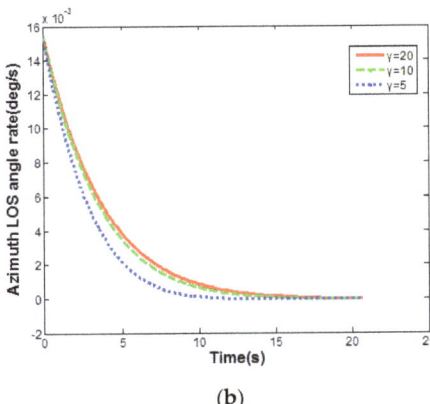

Figure 7. Simulation results in the azimuth plane with different γ. (**a**) Azimuth acceleration. (**b**) Azimuth LOS angle rate.

5. Conclusions

Novel fixed-time convergence guidance laws are proposed for diverse engagement scenarios, and the fixed-time convergence of the LOS angle rate is proven under the proposed laws. The convergence rate is merely related to control parameters, a suitable selection of which can ensure the convergence is fulfilled before the final impact time. Unlike finite-time convergence guidance law, the proposed method is not affected by the chattering effect. The simulation results present high accuracy in a realistic missile system for autopilot first-order time constants as high as 0.5 s. In our future related research, more constraints to improve the missile performance should also be concerned, such as impact time and impact angle.

Author Contributions: Conceptualization, Y.L. and Z.C.; Methodology, Y.L. and Z.C.; Formal analysis, Z.C.; Investigation, C.O., C.S. and Z.C.; Data curation, C.O.; Writing—review & editing, C.S. All authors have read and agreed to the published version of the manuscript.

Funding: This study was co-supported in part by the National Natural Science Foundation of China (No. 61903146).

Data Availability Statement: Not applicable.

Conflicts of Interest: The authors declare no conflict of interest.

References

1. Zarchan, P. *Tactical and Strategic Missile Guidance*, 6th ed.; American Institute of Aeronautics & Astronautics Inc.: Reston, VA, USA, 2012; pp. 569–600.
2. Yang, C.D.; Chen, H.Y. Three-dimensional nonlinear H∞ guidance law. *Int. J. Robust Nonlinear Control* **1999**, *32*, 3079–3084.
3. Liao, F.; Luo, Q.; Ji, H.; Gai, W. Guidance laws with input saturation and nonlinear robust H∞ observers ☆. *ISA Trans* **2016**, *63*, 20–31. [CrossRef] [PubMed]
4. Lechevin, N.; Rabbath, C.A. Lyapunov-Based Nonlinear Missile Guidance. *J. Guid. Control Dyn.* **2004**, *27*, 1096–1102. [CrossRef]
5. Saleem, A.; Ratnoo, A. Lyapunov-Based Guidance Law for Impact Time Control and Simultaneous Arrival. *J. Guid. Control Dyn.* **2015**, *39*, 164–173. [CrossRef]
6. Cheng, Z.; Liu, L.; Wang, Y. Lyapunov-based switched-gain impact angle control guidance. *Chin. J. Aeronaut.* **2018**, *31*, 765–775. [CrossRef]
7. Cheng, Z.; Bo, W.; Lei, L.; Wang, Y. A composite impact-time-control guidance law and simultaneous arrival. *Aerosp. Sci. Technol.* **2018**, *80*, 403–412. [CrossRef]
8. Liu, W.; Wei, Y.; Duan, G. Barrier Lyapunov function-based integrated guidance and control with input saturation and state constraints. *Aerosp. Sci. Technol.* **2019**, *84*, 845–855. [CrossRef]
9. Zhou, D.; Sun, S.; Teo, K.L. Guidance Laws with Finite Time Convergence. *J. Guid. Control Dyn.* **2009**, *32*, 1838–1846. [CrossRef]
10. Zhao, Z.; Li, C.; Yang, J.; Li, S. Output feedback continuous terminal sliding mode guidance law for missile-target interception with autopilot dynamics. *Aerosp. Sci. Technol.* **2019**, *86*, 256–267. [CrossRef]

11. Sun, S.; Zhou, D.; Hou, W.T. A guidance law with finite time convergence accounting for autopilot lag. *Aerosp. Sci. Technol.* **2013**, *25*, 132–137. [CrossRef]
12. Kumar, S.R.; Rao, S.; Ghose, D. Sliding-Mode Guidance and Control for All-Aspect Interceptors with Terminal Angle Constraints. *J. Guid. Control Dyn.* **2012**, *35*, 1230–1246. [CrossRef]
13. Kumar, S.R.; Rao, S.; Ghose, D. Nonsingular Terminal Sliding Mode Guidance with Impact Angle Constraints. *J. Guid. Control. Dyn.* **2014**, *37*, 1114–1130. [CrossRef]
14. Polyakov, A. Nonlinear Feedback Design for Fixed-Time Stabilization of Linear Control Systems. *IEEE Trans. Autom. Control* **2012**, *57*, 2106–2110. [CrossRef]
15. Zhang, Y.; Tang, S.; Guo, J. An adaptive fast fixed-time guidance law with an impact angle constraint for intercepting maneuvering targets. *Chin. J. Aeronaut.* **2018**, *31*, 1327–1344. [CrossRef]
16. Li, G.; Wu, Y.; Xu, P. Fixed-time cooperative guidance law with input delay for simultaneous arrival. *Int. J. Control* **2021**, *94*, 1664–1673. [CrossRef]
17. Zhang, P.; Zhang, X. Multiple missiles fixed-time cooperative guidance without measuring radial velocity for maneuvering targets interception. *ISA Trans.* **2021**, *126*, 388–397. [CrossRef]
18. Chen, Z.; Chen, W.; Liu, X.; Cheng, J. Three-dimensional fixed-time robust cooperative guidance law for simultaneous attack with impact angle constraint. *Aerosp. Sci. Technol.* **2021**, *110*, 106523. [CrossRef]
19. Liu, J.; Zhang, Y.; Liu, H.; Yu, Y.; Sun, C. Robust eventtriggered control of second-order disturbed leader-follower mass: A nonsingular finite-time consensus approach. *Int. J. Robust Nonlinear Control* **2019**, *29*, 4298–4314. [CrossRef]
20. Wang, H.; Zhu, Q.X. Finite-time stabilization of high-order stochastic nonlinear systems in strict-feedback form. *Automatica* **2015**, *54*, 284–291. [CrossRef]
21. Zuo, Z.; Tie, L. A new class of finite-time nonlinear consensus protocols for multi-agent systems. *Int. J. Control* **2014**, *87*, 363–370. [CrossRef]
22. Zuo, Z.; Tie, L. Distributed robust finite-time nonlinear consensus protocols for multi-agent systems. *Int. J. Syst. Sci.* **2016**, *47*, 1366–1375. [CrossRef]
23. Yu, J.; Yu, S.; Li, J.; Yan, Y. Fixed-time stability theorem of stochastic nonlinear systems. *Int. J. Control* **2019**, *92*, 2194–2200 [CrossRef]
24. Li, H.; Li, C.; Huang, T.; Ouyang, D. Fixed-time stability and stabilization of impulsive dynamical systems. *J. Frankl. Inst.-Eng. Appl. Math.* **2017**, *354*, 8626–8644. [CrossRef]
25. Liu, J.; Zhang, Y.; Yu, Y.; Sun, C. Fixed-time event-triggered consensus for nonlinear multiagent systems without continuous communications. *IEEE Trans. Syst. Man Cybern.-Syst.* **2021**, *49*, 2221–2229. [CrossRef]

Disclaimer/Publisher's Note: The statements, opinions and data contained in all publications are solely those of the individual author(s) and contributor(s) and not of MDPI and/or the editor(s). MDPI and/or the editor(s) disclaim responsibility for any injury to people or property resulting from any ideas, methods, instructions or products referred to in the content.

Article

Adaptive Trajectory Tracking Algorithm for The Aerospace Vehicle Based on Improved T-MPSP

Chao Ou [1,2], Chengjun Shan [1], Zhongtao Cheng [1] and Yaosong Long [1,*]

[1] School of Aerospace Engineering, Huazhong University of Science and Technology, Wuhan 430074, China
[2] Aerospace Technology Institute of China Aerodynamics Research and Development Center, Mianyang 621000, China
* Correspondence: longyaosong@hust.edu.cn

Abstract: To deal with the uncertainty and disturbance that exist in the tracking system of an aerospace vehicle, an adaptive trajectory-tracking method based on a novel tracking model predictive static programming (T-MPSP) is proposed. Firstly, to make the proposed method more adaptive to uncertain parameter deviations, an extended Kalman filter (EKF) parameter correction strategy is designed. Then, the control constraints are considered to form a novel T-MPSP algorithm. By combining the parameter correction strategy with the improved T-MPSP algorithm, a novel adaptive tracking guidance scheme is presented. Finally, simulations are carried out to demonstrate the effectiveness of the proposed method.

Keywords: trajectory tracking; uncertainty; extended kalman filter; adaptive; control constraints; T-MPSP

MSC: 90C29

1. Introduction

For the past few decades, aerospace vehicles have been applied in both military and civilian fields, and their performance has been very compelling. For missions involving aerospace vehicles, a good trajectory-tracking ability is the essential prerequisite for the successful application of the vehicles [1]. In a traditional trajectory-tracking process for an aerospace vehicle, the dynamic model of the vehicle and the desired trajectory satisfying all the constraints are given in advance. Then, the control methods are designed to guide the vehicle to track the desired trajectory. However, the trajectories of aerospace vehicles usually cover a wide range of altitudes, which may lead to a dramatic change in the atmospheric environment during the trajectory-tracking process. In addition to the disturbance brought about by the dramatic change in the external environment, there are also uncertainties about the dynamic system of the vehicles [2,3]. Thus, the precise model information is not always available, and the traditional trajectory-tracking method cannot achieve satisfactory performance under these circumstances. As a result, the design of the novel trajectory-tracking control method to cope with disturbance and uncertainty is very important.

One of the most commonly used control architectures for trajectory-tracking control is the proportional-integral-derivative (PID) controller and its variants [4–7], which are known for the simplicity of their framework. To improve the tracking performance, many modern control theories, including intelligent optimization [4], feedback control [5], and fuzzy control [6,7], have been applied to the trajectory-tracking method design process. Another popular control architecture for trajectory-tracking control derives from the sliding mode control theory. In [8], the sliding mode control theory is applied to make the tracking error converge to zero in a finite time period. As an improvement of the method in [8], the fixed-time control theory was combined with the sliding mode control theory. The robustness of the resulting tracking laws was further enhanced [9,10]. Recently, the neural

network and adaptive updating laws were applied to the sliding mode control framework. The convergence of the tracking error was ensured by the sliding mode control approach and any system uncertainties were handled by the neural network [11].

In addition, the nonlinear model predictive control (NMPC) was applied to the trajectory-tracking control [12–15]. However, several issues exist in the implementation of the NMPC-based tracking methods, which are summarized as follows: (1) a universal and exact model of the whole tracking system is difficult to acquire, (2) the calculation speed of the NMPC cannot meet the requirement in practice. A feasible way to cope with these issues is to use model predictive static programming (MPSP) [16], which combines the characteristics of approximate dynamic programming [17] and NMPC. The MPSP has been proven to be an effective method to cope with two-point boundary value problems with terminal constraints, which has significant advantages: (1) the terminal constraints are transformed into linear equality constraints with only the control variable to be optimized, (2) the closed analytical expression of the appropriate objective function is acquired from the static algorithm, (3) the sensitivity matrix of the algorithm can be calculated skillfully using recursion, which improves the calculation speed. The MPSP method has been used in many applications, such as in the trajectory control of launch vehicles, reentry guidance, cooperative control, etc. [18–20]. Typically, the MPSP method aims to improve the terminal accuracy by iteratively discretizing and updating. For a trajectory-tracking problem, only the terminal-tracking accuracy can be guaranteed using the MPSP method, and its computing time increases exponentially with the number of discrete nodes that are tracked. To deal with this drawback, the T-MPSP is put forward to track a trajectory over a receding horizon window. Meanwhile, the predicted time horizon can be set manually to balance the computational efficiency and the precision, which is another advantage of the T-MPSP algorithm [21,22].

Some trajectory-tracking methods are based directly on the input and output data of the system. For example, a robust model-free controller for trajectory-tracking control is proposed in Ref. [23]. No dynamic model information of the controlled system is needed, and the resulting controller is a combination of the PID controller and the sliding mode control. As an improvement of the work in Ref. [23], a forecasting-based data-driven model-free adaptive sliding mode attitude control method is proposed for the post-capture combined spacecraft with unknown inertial properties and external disturbances [24]. To cope with the unknown dynamics of the system, a model-free control method is proposed via the time-varying compensation of the un-modeled system [25]. An iterative sliding mode control technique is utilized to design the adaptive model-free trajectory-tracking method in Ref. [26]. Although these model-free control methods can deal with model uncertainties, the operation data require a huge memory size, and the computing burden is heavy.

Hence, considering the uncertainty and disturbance that exist in the tracking system of the aerospace vehicle, an adaptive trajectory-tracking method based on the novel T-MPSP is proposed. Firstly, to cope with the uncertain parameter deviations, an EKF parameter correction strategy is designed. Then, the control constraints are considered to form a novel T-MPSP algorithm. By combining the parameter correction strategy with the improved T-MPSP algorithm, an adaptive tracking guidance scheme is presented. Finally, simulations are carried out under various deviation conditions to verify the reliability and robustness of the proposed method.

The main contributions of this paper are summarized as follows:

(1) To our best knowledge, no existing methods have applied the EKF with the T-MPSP to solve the trajectory problems of aerospace vehicles.
(2) Compared with the MPSP method in [18–20], the proposed method has a fast computing speed and high accuracy.
(3) Compared with [21–23], the proposed scheme can cope with the control constraints.

The rest of this paper is organized as follows: Section 2 presents the model of the aerospace vehicle. Section 3 presents the online parameter identification method and the improved T-MPSP algorithm. The simulations are presented in Section 4. Finally, the conclusion is presented in Section 5.

2. Model of the Aerospace Vehicle

In this section, first, the dynamic model describing the motion of the aerospace vehicle is presented. Then, the constraints that should be satisfied are introduced.

2.1. Dynamic Equations

The three-dimensional mass point dynamic equations [22,27] of the aircraft in the longitudinal plane are

$$\begin{cases} \frac{dh}{dt} = v \sin \gamma \\ \frac{dv}{dt} = \frac{T \cos \alpha - D}{m} - g \sin \gamma \\ \frac{d\gamma}{dt} = \frac{T \sin \alpha + L}{mv} - \frac{g}{v} \cos \gamma + \frac{v}{r} \cos \gamma \\ \frac{dm}{dt} = -\frac{T}{g_0 I_{sp}} \end{cases} \quad (1)$$

where h, v, γ, α and m represent the altitude, velocity, flight path angle, angle of attack (AOA), and mass, respectively. g and g_0 represent the gravitational acceleration at the current altitude and on the earth's surface, respectively. I_{sp} represents the specific impulse of the engine. T, L and D denote the engine thrust, aerodynamic lift, and drag, respectively, which are defined as

$$\begin{cases} T = 0.029 \phi I_{sp} \rho g_0 v C_T A_C \\ L = \frac{1}{2} \rho v^2 S_{ref} C_L(\text{Ma}, \alpha) \\ D = \frac{1}{2} \rho v^2 S_{ref} C_D(\text{Ma}, \alpha) \end{cases} \quad (2)$$

where C_T, C_L, and C_D denote the thrust, lift and drag coefficients, respectively. ϕ denotes the throttle. ρ, S_{ref} and Ma denote the atmospheric density, reference area, and Mach number, respectively.

2.2. Flight Constraints

To ensure flight safety, the dynamic pressure constraint is considered, which is defined as

$$q = \frac{1}{2} \rho v^2 \leq q_{max} \quad (3)$$

In addition, the aircraft must satisfy the terminal constraints, which are described as

$$\begin{cases} \left| h_f - h_f^* \right| \leq \varepsilon_h \\ \left| v_f - v_f^* \right| \leq \varepsilon_v \\ \left| \gamma_f - \gamma_f^* \right| \leq \varepsilon_\gamma \end{cases} \quad (4)$$

where the subscript f refers to the final value and the superscript $*$ refers to the desired value. ε_h, ε_v, and ε_γ represent the deviation thresholds of the terminal altitude, velocity, and flight path angle, respectively.

The control variables considered are α and the throttle ϕ, the constraints of which are as follows:

$$\begin{cases} \alpha_{min} \leq \alpha \leq \alpha_{max} \\ \phi_{min} \leq \phi \leq \phi_{max} \end{cases} \quad (5)$$

Additionally, to prevent the control variable from changing too drastically, the rate of AOA must satisfy the following constraint:

$$\dot{\alpha}_{min} \leq \dot{\alpha} \leq \dot{\alpha}_{max} \quad (6)$$

3. The Trajectory-Tracking Strategy

In this section, we will introduce an online parameter identification method and an improved T-MPSP algorithm. By combining these two methods, an adaptive trajectory tracking guidance algorithm based on an improved T-MPSP will be presented.

3.1. Online Parameter Identification Method

Since an accurate model can improve the trajectory-tracking accuracy, a parameter correction strategy based on the EKF is designed in this subsection. In the actual tracking process, there are deviations in the parameters of atmospheric density, thrust coefficient, lift coefficient, and drag coefficient, which will make the model inaccurate. We define these parameters as

$$\rho = \rho^*(1+\Delta\rho), C_T = C_T^*(1+\Delta C_T), C_L = C_L^*(1+\Delta C_L), C_D = C_D^*(1+\Delta C_D) \quad (7)$$

where ρ^*, C_T^*, C_L^* and C_D^* are the desired values. ρ, C_T, C_L and C_D are the actual values. $\Delta\rho$, ΔC_T, ΔC_L, and ΔC_D respectively represent the unknown deviation of each parameter. These unknown deviations are written in an overall form as

$$\beta = [\Delta\rho\ \Delta C_T\ \Delta C_L\ \Delta C_D]^T \quad (8)$$

Defining $x = [h\ v\ \gamma\ m]$, the following dynamic equations are obtained from (1) and (7) as

$$\dot{x} = f(x, u, \beta) \quad (9)$$

in which

$$f(x,u,\beta) = \begin{bmatrix} v\sin\gamma \\ \frac{T^*(1+\Delta\rho)(1+\Delta C_T)\cos\alpha - D^*(1+\Delta\rho)(1+\Delta C_D)}{m} - g\sin\gamma \\ \frac{T^*(1+\Delta\rho)(1+\Delta C_T)\sin\alpha + L^*(1+\Delta\rho)(1+\Delta C_L)}{mv} - \frac{g}{v}\cos\gamma + \frac{v}{r}\cos\gamma \\ -\frac{T^*(1+\Delta\rho)(1+\Delta C_T)}{g_0 I_{sp}} \end{bmatrix} \quad (10)$$

Since the EKF algorithm [28] needs to augment the unknown parameters into the state variables of the system, the new augmented dynamic equation is defined as

$$\dot{x}_a = f_a(x_a, u) + \omega_a = \begin{bmatrix} f(x,u,\beta) \\ 0 \end{bmatrix} + \omega_a \quad (11)$$

where

$$\begin{cases} x_a = [h\ v\ \gamma\ m\ \Delta\rho\ \Delta C_T\ \Delta C_L\ \Delta C_D]^T \\ u = [\alpha\ \phi]^T \end{cases} \quad (12)$$

ω_a represents the uncorrelated zero-mean white Gaussian noise. To identify the unknown parameters, the EKF algorithm also requires the measurement information of the unknown parameters. The measurement equation is expressed as

$$y = \begin{bmatrix} x_m \\ a_{xm} \\ a_{zm} \\ q_m \end{bmatrix} = h(x_a, u) + v_a = \begin{bmatrix} x \\ \frac{\rho v^2}{2m}C_x + \frac{T}{m} \\ \frac{\rho v^2}{2m}C_z \\ \frac{\rho v^2}{2} \end{bmatrix} + v_a \quad (13)$$

where x_m, a_{xm}, a_{zm} and q_m denote the state values, axial acceleration, normal acceleration and dynamic pressure measured by the sensor, respectively. v_a represents the uncorrelated zero-mean Gaussian white noise. C_x and C_z refer to axial and normal force coefficients, respectively. The calculation formulas of C_x and C_z are

$$\begin{cases} C_x = C_L \sin \alpha - C_D \cos \alpha \\ C_z = -C_L \cos \alpha - C_D \sin \alpha \end{cases} \quad (14)$$

Then, the two-step online parameter identification method will be introduced as follows:

a. Prediction

First, the prior estimate of the state at the current time instant k is

$$\hat{x}_{a_k}^- = \hat{x}_{a_{k-1}} + f(\hat{x}_{a_{k-1}})\Delta t + \frac{F_k f(\hat{x}_{a_{k-1}})}{2}\Delta t^2 \quad (15)$$

where F_k represents the Jacobian matrix of the augmented state equation $f_a(\hat{x}_{a_{k-1}})$ to estimate the state $\hat{x}_{a_{k-1}}$ at the previous time instant and Δt represents the sampling time interval. Then, the error covariance matrix of the current time instant k according to the equation is acquired as

$$P_k^- = \Phi_k P_{k-1} \Phi_k^T + Q_k \quad (16)$$

where $\Phi_k = I + F_k \Delta t$ represents the state transition matrix, and Q_k represents the noise covariance matrix.

b. Update

Using the error covariance matrix P_k^-, we update the Kalman filter gain coefficient K_k at the current time instant k as

$$K_k = P_k^- H_k^T \left(H_k P_k^- H_k^T + V_k \right)^{-1} \quad (17)$$

where V_k represents the measurement noise covariance matrix, and H_k represents the Jacobian matrix of the prior estimate $\hat{x}_{a_k}^-$ of the state from the measurement equation $y(\hat{x}_{a_k}^-)$. Subsequently, the error covariance matrix is updated as

$$P_k = (I - K_k H_k) P_k^- (I - K_k H_k)^T + K_k R_k K_k^T \quad (18)$$

The measurement correction is updated as

$$\Delta y_k = y_k - y(\hat{x}_{a_k}^-) \quad (19)$$

where y_k represents the actual measurement value.

Finally, we update the posterior estimate of the state at the current time instant k using

$$\hat{x}_{a_k} = \hat{x}_{a_k}^- + K_k \Delta y_k \quad (20)$$

$\hat{x}_{a_k}$ of the augmented state has been obtained through the EKF algorithm, from which the estimated value of the corresponding unknown parameter β is also obtained. Thus, the new model that accounts for uncertain derivations is obtained by substituting the estimated value of β into (10).

Remark 1. An imprecise model will make it difficult for the trajectory-tracking method design, and uncertain parameter deviations will affect the accuracy of the model. Hence, a parameter correction strategy based on the EKF is designed in this subsection. By applying the EKF algorithm, the augmented state $\hat{x}_{a_k}$ is obtained, from which the estimated value of the corresponding unknown parameter β is also obtained. Then, the exact model that considers uncertain derivations is obtained.

3.2. Improved T-MPSP Algorithm

According to (1), the state variables are $X = [h, v, \gamma, m]^T$, the control variables are $U = [\alpha, \phi]^T$, and the output variables are $Y = [h, v, \gamma]^T$. The discrete form of the state equation and output equation of a continuous nonlinear system is expressed as

$$X_{k+1}^i = F_k\left(X_k^i, U_k^i\right) \tag{21}$$

$$Y_k^i = h\left(X_k^i\right) \tag{22}$$

where $X \in \mathbb{R}^n$, $U \in \mathbb{R}^m$ and $Y \in \mathbb{R}^p$ denote the state, control, and output variables, respectively. $k = 1, 2, \cdots, N_y$ represents the kth sampling point, and N_y represents the total length of the prediction time domain. $k = 1$ and $k = N_y$ represent the starting and ending points of the prediction time domain, respectively, and i represents the number of iterations.

The main objective of the trajectory-tracking algorithm is to find the suitable control variables U_k^{i+1} to make the output Y_k^{i+1} as close to the desired output $Y_k^* (k = 2, 3, \cdots, N_y)$ as possible at each sampling time; that is, $Y_k^{i+1} \to Y_k^*$, where U_k^i represents the control variables at the current ith iteration, and U_k^{i+1} represents the control variables for the next iteration.

Similarly, Y_k^i is the current output, and Y_k^{i+1} is the output of the next iteration. We should also add a performance index that minimizes the deviation of the control variables to avoid overly drastic changes in the control variables. Therefore, the following objective function is proposed:

$$J = \frac{1}{2}\sum_{k=2}^{N_y} \left(Y_k^{i+1} - Y_k^*\right)^T Q_k^i \left(Y_k^{i+1} - Y_k^*\right) + \frac{1}{2}\sum_{k=1}^{N_y-1} \left(U_k^{i+1} - U_k^i\right)^T R_k^i \left(U_k^{i+1} - U_k^i\right) \tag{23}$$

where Q_k^i and R_k^i are both the positive definite weight matrix at the ith iteration.

The deviation vectors between two consecutive iterations at the same time are defined as follows:

$$Y_k^{i+1} = Y_k^i + \Delta Y_k^i \tag{24}$$

$$X_k^{i+1} = X_k^i + \Delta X_k^i \tag{25}$$

$$U_k^{i+1} = U_k^i + \Delta U_k^i \tag{26}$$

By expanding the Taylor series of ΔY_k^i and ignoring its higher-order terms, we obtain

$$\Delta Y_k^i \approx dY_k^i = \left[\frac{\partial Y_k}{\partial X_k}\right] dX_k^i \tag{27}$$

Similarly, we obtain

$$\Delta X_{k+1}^i \approx dX_{k+1}^i = \left[\frac{\partial F_k}{\partial X_k}\right] dX_k^i + \left[\frac{\partial F_k}{\partial U_k}\right] dU_k^i \tag{28}$$

where dX_k^i and dU_k^i represent deviations in the state variables and control variables at the kth time instant, respectively. By substituting (27) into (26), we obtain

$$dY_k^i = \left[\frac{\partial Y_k}{\partial X_k}\right]\left[\frac{\partial F_{k-1}}{\partial X_{k-1}}\right] dX_{k-1}^i + \left[\frac{\partial Y_k}{\partial X_k}\right]\left[\frac{\partial F_{k-1}}{\partial U_{k-1}}\right] dU_{k-1}^i \tag{29}$$

where dX_{k-1} is also expanded into an equation composed of dX_{k-2} and dU_{k-2}, and the corresponding equation is substituted into (29). By successively substituting $dX_k, dX_{k-1}\ldots, dX_2$ into the expression of dY_k^i, we obtain

$$dY_k^i = \left[A^k\right]^i dX_1^i + \left[B_1^k\right]^i dU_1^i + \left[B_2^k\right]^i dU_2^i + \cdots + \left[B_{k-1}^k\right]^i dU_{k-1}^i \tag{30}$$

It is obvious that $dX_1^i = 0$. Then, (30) is simplified to

$$dY_k^i = \sum_{j=1}^{k-1} \left[B_j^k\right]^i dU_j^i \tag{31}$$

$$\left[B_j^k\right]^i = \left[\frac{\partial Y_k}{\partial X_k}\right]\left[\frac{\partial F_{k-1}}{\partial X_{k-1}}\right]\cdots\left[\frac{\partial F_{j+1}}{\partial X_{j+1}}\right]\left[\frac{\partial F_j}{\partial U_j}\right], k = 2, 3, \cdots, N_y \tag{32}$$

where $\left[B_j^k\right]^i$ is called the sensitivity matrix. Then, the deviations in the output variables and control variables at each sampling time are formed into a linear equation, in which each dU_j^i is a variable to be optimized. (32) is calculated recursively in the following way:

$$\begin{aligned}
\left[A_k^k\right]^i &= I_{n\times n} \\
\left[A_j^k\right]^i &= \left[A_{j+1}^k\right]^i \left[\frac{\partial F_j}{\partial X_j}\right] \\
\left[B_j^k\right]^i &= \left[\frac{\partial Y_k}{\partial X_k}\right]\left[A_{j+1}^k\right]^i \left[\frac{\partial F_j}{\partial U_j}\right]
\end{aligned} \tag{33}$$

where $k = 2, 3, \ldots, N_y, j = (k-1), (k-2), \ldots, 1$, when $j \geq k$, $\left[B_j^k\right]^i = 0_{p\times m}$.

According to (24) and (26), and considering the small approximation $\Delta Y_k^i \approx dY_k^i$, $\Delta U_k^i \approx dU_k^i$, we obtain

$$\begin{cases} \left(Y_k^{i+1} - Y_k^*\right) = \Delta Y_k^i + \left(Y_k^i - Y_k^*\right) = \Delta Y_k^i - \Delta Y_k^{*i} = dY_k^i - \Delta Y_k^{*i} \\ \left(U_k^{i+1} - U_k^i\right) = \Delta U_k^i = dU_k^i \end{cases} \tag{34}$$

Hence, the objective function in (23) is rewritten as

$$J = \frac{1}{2}\sum_{k=2}^{N_y} \left(dY_k^i - \Delta Y_k^{*i}\right)^T Q_k^i \left(dY_k^i - \Delta Y_k^{*i}\right) + \frac{1}{2}\sum_{k=1}^{N_y-1} \left(dU_k^i\right)^T R_k^i \left(dU_k^i\right) \tag{35}$$

In the traditional T-MPSP algorithm, the increment of the control variables is added to the performance index to indirectly constrain the control variables. However, this method cannot strictly constrain the control variables, which may fail to satisfy the constraints. Therefore, we propose an improved T-MPSP algorithm by adding control variable constraints.

The control variable constraints (5) and (6) are expressed as

$$\begin{aligned}
U_{\min} &\leq U_k^{i+1} \leq U_{\max} \\
\dot{U}_{\min} &\leq \dot{U}_k^{i+1} = \frac{U_{k+1}^{i+1} - U_k^{i+1}}{h} \leq \dot{U}_{\max}
\end{aligned} \tag{36}$$

where h represents the sampling step size, $U \in \mathbb{R}^m$. According to (34), we obtain

$$U_{\min} \leq U_k^i + dU_k^i \leq U_{\max} \Rightarrow \begin{cases} dU_k^i \leq U_{\max} - U_k^i \\ -dU_k^i \leq U_k^i - U_{\min} \end{cases} \tag{37}$$

which is simplified to

$$\begin{cases} dU_k^i \leq U_{\max} - U_k^i = C1_k^i \\ -dU_k^i \leq U_k^i - U_{\min} = C2_k^i \end{cases} \quad (38)$$

The control magnitude constraint is thus transformed into a linear inequality constraint with the only unknown dU_k^i. Then, $\dot{U}_k^{i+1}$ is converted into

$$\dot{U}_k^{i+1} = \frac{U_{k+1}^{i+1} - U_k^{i+1}}{h} = \frac{(U_{k+1}^i + dU_{k+1}^i) - (U_k^i + dU_k^i)}{h} \quad (39)$$

Thereby,

$$\dot{U}_{\min} \leq \frac{(U_{k+1}^i + dU_{k+1}^i) - (U_k^i + dU_k^i)}{h} \leq \dot{U}_{\max} \Rightarrow \begin{cases} -dU_k^i + dU_{k+1}^i \leq h\dot{U}_{\max} - U_{k+1}^i + U_k^i \\ dU_k^i - dU_{k+1}^i \leq -h\dot{U}_{\min} + U_{k+1}^i - U_k^i \end{cases} \quad (40)$$

which is simplified to

$$\begin{cases} \begin{bmatrix} -I_m & I_m \end{bmatrix} \begin{bmatrix} dU_k^i \\ dU_{k+1}^i \end{bmatrix} \leq h\dot{U}_{\max} - U_{k+1}^i + U_k^i = C3_k^i \\ \begin{bmatrix} I_m & -I_m \end{bmatrix} \begin{bmatrix} dU_k^i \\ dU_{k+1}^i \end{bmatrix} \leq -h\dot{U}_{\min} + U_{k+1}^i - U_k^i = C4_k^i \end{cases} \quad (41)$$

where I_m is an identity matrix of order $m \times m$. In this way, the change rate constraint is transformed into linear inequality constraints with unknowns dU_k^i and dU_{k+1}^i. In order to facilitate the subsequent solution, (41) is rewritten in the following form:

$$\begin{cases} PdU^i = C3^i \\ -PdU^i = C4^i \end{cases} \quad (42)$$

where $P = \begin{bmatrix} -I_m & I_m & 0 & \cdots & 0 \\ 0 & -I_m & I_m & \cdots & 0 \\ 0 & 0 & \ddots & \ddots & 0 \\ 0 & 0 & 0 & -I_m & I_m \end{bmatrix}$, $dU^i = \begin{bmatrix} dU_1^i \\ dU_2^i \\ \vdots \\ dU_{N-2}^i \\ dU_{N-1}^i \end{bmatrix}$, $C3^i = \begin{bmatrix} C3_1^i \\ C3_2^i \\ \vdots \\ C3_{N-2}^i \\ C3_{N-1}^i \end{bmatrix}$, $C4^i = \begin{bmatrix} C4_1^i \\ C4_2^i \\ \vdots \\ C4_{N-2}^i \\ C4_{N-1}^i \end{bmatrix}$.

Therefore, combining (35) with the inequality constraints in (38) and (42) yields

$$\begin{aligned} \min J &= \tfrac{1}{2} \sum_{k=2}^{N_y} \left(dY_k^i - \Delta Y_k^{*i} \right)^T Q_k^i \left(dY_k^i - \Delta Y_k^{*i} \right) + \tfrac{1}{2} \sum_{k=1}^{N_y-1} \left(dU_k^i \right)^T R_k^i \left(dU_k^i \right) \\ \text{s.t.} & \begin{cases} dU_k^i \leq C1_k^i \\ -dU_k^i \leq C2_k^i \end{cases}, \begin{cases} PdU^i = C3^i \\ -PdU^i = C4^i \end{cases} \end{aligned} \quad (43)$$

In this paper, the Lagrangian multiplier method and the penalty function are applied to solve the NLP problem; thus, we have

$$\begin{aligned}
L &= \tfrac{1}{2}\sum_{k=2}^{N_y}\left(\sum_{j=1}^{k-1} B_j^k dU_j - \Delta Y_k^*\right)^T Q_k \left(\sum_{j=1}^{k-1} B_j^k dU_j - \Delta Y_k^*\right) \\
&+ \tfrac{1}{2}\sum_{k=1}^{N_y-1}(dU_k)^T R_k(dU_k) \\
&+ \tfrac{1}{2}\sum_{k=1}^{N_y-1}(dU_k - C1_k)^T \sigma 1_k (dU_k - C1_k) \\
&+ \tfrac{1}{2}\sum_{k=1}^{N_y-1}(-dU_k - C2_k)^T \sigma 2_k (-dU_k - C2_k) \\
&+ \tfrac{1}{2}\sum_{k=1}^{N_y-2}\left(\sum_{j=1}^{N_y-1} P_{kj} dU_j - C3_k\right)^T \sigma 3_k \left(\sum_{j=1}^{N_y-1} P_{kj} dU_j - C3_k\right) \\
&+ \tfrac{1}{2}\sum_{k=1}^{N_y-2}\left(\sum_{j=1}^{N_y-1} -P_{kj} dU_j - C4_k\right)^T \sigma 4_k \left(\sum_{j=1}^{N_y-1} -P_{kj} dU_j - C4_k\right)
\end{aligned} \tag{44}$$

where P_{kj} represents the element of the kth row and the jth column of the matrix P. The data in the following equations are all in the same iteration i, so the superscript i is omitted for the convenience of subsequent derivation. According to the necessary conditions for first-order optimality, we obtain

$$\frac{\partial L}{\partial dU_l} = 0$$

$$\begin{aligned}
\Rightarrow R_l dU_l &+ \sum_{k=2}^{N_y}\left(\left(B_l^k\right)^T Q_k \sum_{j=1}^{k-1} B_j^k dU_j\right) - \sum_{k=2}^{N_y}\left(B_l^k\right)^T Q_k \Delta Y_k^* + \sigma 1_l(dU_l - C1_l) + \\
\sigma 2_l(dU_l + C2_l) &+ \sum_{k=2}^{N_y-2}\left((P_{kl})^T \sigma 3_k \sum_{j=1}^{N_y-1} P_{kj} dU_j\right) - \sum_{k=2}^{N_y-2}(P_{kl})^T \sigma 3_k C3_k + \\
\sum_{k=2}^{N_y-2}&\left((P_{kl})^T \sigma 4_k \sum_{j=1}^{N_y-1} P_{kj} dU_j\right) + \sum_{k=2}^{N_y-2}(P_{kl})^T \sigma 4_k C4_k = 0
\end{aligned} \tag{45}$$

where $\sigma 1_k$, $\sigma 2_k$, $\sigma 3_k$ and $\sigma 4_k$ are all penalty factors. By changing the positions of the terms in Equation (45), we obtain

$$\begin{aligned}
(R_l + \sigma 1_l + \sigma 2_l)dU_l &+ \sum_{k=2}^{N_y}\left(\left(B_l^k\right)^T Q_k \sum_{j=1}^{k-1} B_j^k dU_j\right) + \\
\sum_{k=2}^{N_y-2}&\left((P_{kl})^T \left(\sigma 3_k \sum_{j=1}^{N_y-1} P_{kj} dU_j + \sigma 4_k \sum_{j=1}^{N_y-1} P_{kj} dU_j\right)\right) \\
&= \sum_{k=2}^{N_y}\left(B_l^k\right)^T Q_k \Delta Y_k^* + \sum_{k=2}^{N_y-2} P_{kl}^T(\sigma 3_k C3_k - \sigma 4_k C4_k) + \sigma 1_l C1_l - \sigma 2_l C2_l
\end{aligned} \tag{46}$$

All $dU_l (l = 1, 2, \ldots, N_y - 1)$ in (46) are expressed in a matrix form as

$$\begin{bmatrix} M_{11} + T_1 & \cdots & M_{1(N_y-2)} & M_{1(N_y-1)} \\ M_{21} & M_{22} + T_2 & \cdots & M_{2(N-1)} \\ \vdots & \cdots & \ddots & \vdots \\ M_{(N_y-1)1} & \cdots & \cdots & M_{(N_y-1)(N_y-1)} + T_{N_y-1} \end{bmatrix} \begin{bmatrix} dU_1 \\ dU_2 \\ \vdots \\ dU_{N_y-1} \end{bmatrix} = \begin{bmatrix} b_1 \\ b_2 \\ \vdots \\ b_{N_y-1} \end{bmatrix} \tag{47}$$

where

$$\begin{cases} M_{ij} = \sum_{l=(j+1)}^{N_y} B_i^{lT} Q_l B_j^l + \sum_{k=2}^{N_y-2}\left(P_{ki}^T \sigma 3_k P_{kj} + P_{ki}^T \sigma 4_k P_{kj}\right) \\ b_i = \sigma 1_i C1_i - \sigma 2_i C2_i + \sum_{l=2}^{N_y} B_i^{lT} Q_l \Delta Y_l^* + \sum_{k=2}^{N_y-2} P_{ki}^T(\sigma 3_k C3_k - \sigma 4_k C4_k) \\ T_i = R_i + \sigma 1_i + \sigma 2_i \end{cases} \tag{48}$$

By solving (47), we obtain the corrections for all control variables $dU = \begin{bmatrix} dU_1, dU_2, \ldots, dU_{N_y-1} \end{bmatrix}$. Finally, the updated control variables are calculated as

$$U^{i+1} = U^i + dU \tag{49}$$

3.3. Overall Structure and Operating Steps

In this subsection, we combine the online parameter identification method with the improved T-MPSP algorithm to propose an adaptive trajectorytracking guidance algorithm based on an improved T-MPSP. By applying the first-order Euler method, we obtain

$$X_{k+1} = X_k + hf(X, U) = F_k(X_k, U_k) \tag{50}$$

where h represents the sampling step size. $X_k = [h_k, v_k, \gamma_k, m_k]^T$, $U_k = [\alpha_k, \phi_k]^T$ and $Y_k = [h_k, v_k, \gamma_k]^T$ denote the state, control, and output variables of the aircraft, respectively. $f(X, U)$ represents the state differential equation of (1). The comprehensive block diagram of the proposed method is presented in Figure 1, and the operating steps of the adaptive tracking guidance algorithm based on the improved T-MPSP are as follows:

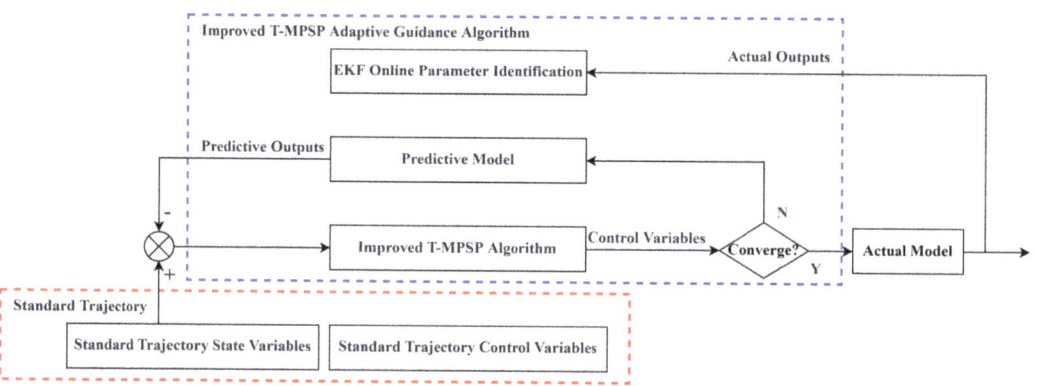

Figure 1. The block diagram of the proposed method.

Step 1: Parameter initialization, such as initializing the prediction time domain, sampling step size, EKF parameters, etc.

Step 2: Use the EKF algorithm to identify the parameter deviation online, which is used to correct the prediction model for the T-MPSP algorithm.

Step 3: In the improved T-MPSP algorithm, use the revised model to update the control variables until they converge or the algorithm reaches the maximum number of iterations.

Step 4: Use the first control values of the prediction time-domain window of the T-MPSP algorithm as the control variables of the current time.

Step 5: Integrate the dynamic equations using fourth-order Runge-Kutta to the next time instant.

Step 6: Determine whether the terminal time is reached. If it is reached, stop the operation. Otherwise, go to step 2 and continue with the same steps.

4. Simulations

In this section, simulations are carried out to show the effectiveness of the proposed method. First, comparison simulations against the existing T-MPSP tracking method are presented. Then, several extreme combinations of parameter deviations are considered in the simulation. Finally, the Monte Carlo simulations are carried out to verify the robustness of the proposed method.

The control parameters are initialized as

$$\begin{cases} N_y = 7 \\ Q_k = \begin{bmatrix} 1.11 & 0 & 0 \\ 0 & 1.96 & 0 \\ 0 & 0 & 5000 \end{bmatrix} \\ R_k = \begin{bmatrix} 1000 \\ 90000 \end{bmatrix} \end{cases} \tag{51}$$

And the constraints considered in this simulation are tabulated in Table 1.

Table 1. Constraints.

Process constraints	q_{max}	150 (kPa)
Control constraints	$[\alpha_{min}, \alpha_{max}]$	$[-3, 21]$ (°)
	$[\phi_{min}, \phi_{max}]$	$[0, 2]$
	$[\dot{\alpha}_{min}, \dot{\alpha}_{max}]$	$[-5, 5]$ (°/s)
Terminal constraint	ε_h	500 (m)
	ε_v	50 (m/s)
	ε_γ	0.5 (°)

4.1. Comparison Simulations

To demonstrate the effectiveness of the proposed method, both the open-loop tracking method and the MPSP tracking law presented in [22] are considered as comparison methods in this subsection. The detailed simulation results are presented in Table 2 and Figure 1.

Table 2. Comparison results.

	Terminal Height Deviations (m)	Terminal Velocity Deviations (m/s)	Terminal Flight Path Angle Deviations (°)
Proposed method	32.95	−12.53	0.279
Open-loop tracking method	−444.49	272.40	1.124
T-MPSP method	−242.36	−15.48	0.025

In Figure 2, the solid lines in blue represent the tracking results of the proposed method, dash lines in blue refer to the tracking results of the open-loop control, dot lines in pink denote the tracking results of the comparison T-MPSP method, and the reference curves are presented in dash-dot form.

For the open-loop control scenario, the tracking performance is far from satisfactory. Moreover, the maximum dynamic pressure violates the constraint, which will cause damage to the vehicle. As for the comparison T-MPSP method, the terminal accuracy is acceptable, while the overall tracking performance is not so good. In the proposed method, the terminal deviations are much smaller than in the comparison law, which confirms the effectiveness and superiority of the proposed method. The optimization process of the proposed method and the T-MPSP method requires about 1.7 s on our laptop with an AMD 1.8 GHz CPU. However, it needs about 2.8 s for the MPSP tracking method.

4.2. The Monte Carlo Simulations

In this subsection, trajectory tracking is carried out under 300 sets of deviations sampled through the LHS method. Furthermore, the distribution intervals of the four uncertain conditions are as follows:

$$\begin{cases} \Delta\rho \in [-0.1, 0.1] \\ \Delta C_T \in [-0.05, 0.05] \\ \Delta C_L \in [-0.1, 0.1] \\ \Delta C_D \in [-0.1, 0.1] \end{cases} \quad (52)$$

The terminal state deviations under the 300 groups of random deviations are presented in Figures 3–5. The statistical maximum deviations in the terminal states are recorded in Table 3.

Table 3. Maximum tracking deviations.

Terminal Height Deviations (m)	Terminal Velocity Deviations (m/s)	Terminal Flight Path Angle Deviations (°)	Dynamic Pressure (kPa)
148.73	47.26	0.313	142.12

(a) Height

(b) Velocity

(c) Flight path angle

(d) AOA

(e) Throttle

(f) Dynamic pressure

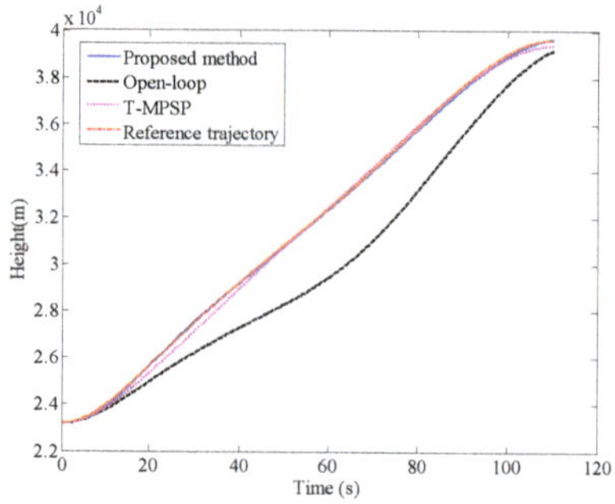

Figure 2. Comparison results.

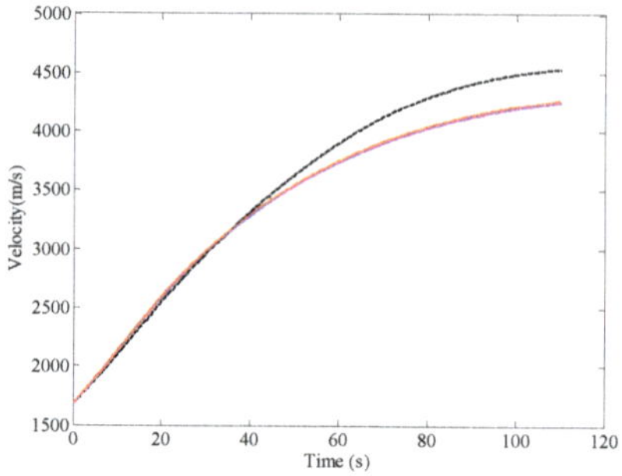

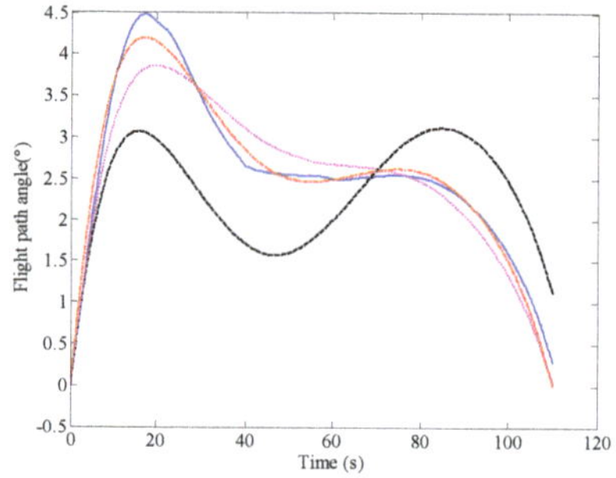

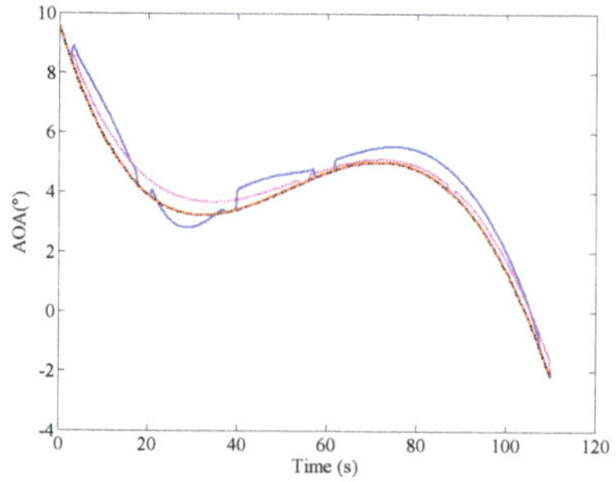

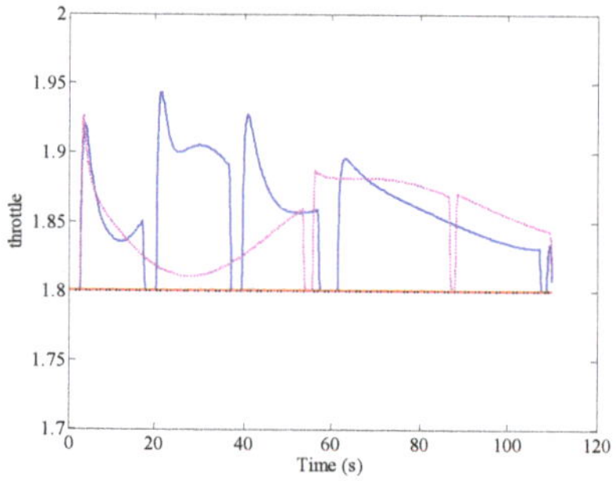

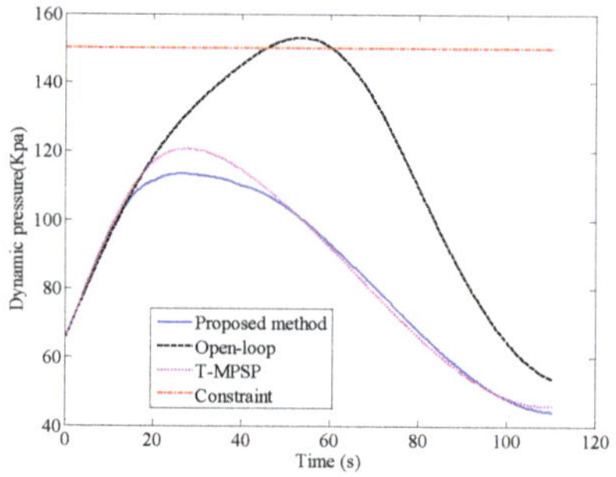

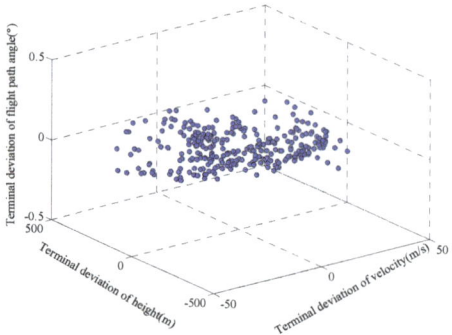

Figure 3. Terminal state deviation distributions.

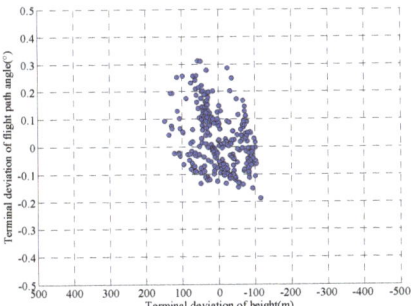

Figure 4. Terminal state deviation distributions (left view).

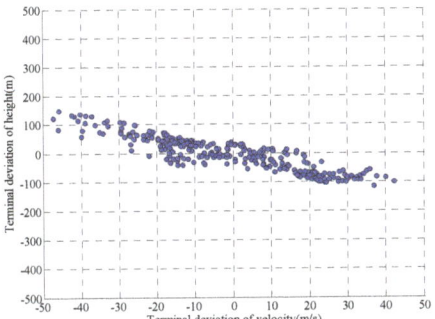

Figure 5. Terminal state deviation distributions (vertical view).

The maximum tracking deviation in the terminal height is less than 150 m, which is far lower than the requirements of terminal height accuracy. At the same time, the maximum tracking deviation in the terminal speed is less than 50 m/s, which also meets the requirements. In addition, the absolute value of terminal flight path angle deviation is also within the allowable range of 0.5°, and the maximum dynamic pressure also meets the constraints. These simulation results show that the novel T-MPSP method proposed in this paper has a good anti-interference ability against coefficient deviations caused by complex flight environments.

5. Conclusions

An adaptive trajectory-tracking method based on a novel tracking model predictive static programming (T-MPSP) was proposed. The proposed method had advantages in computing efficiency and tracking accuracy. Further, it was more adaptive to uncertain parameter deviations due to the parameter correction strategy. Firstly, a parameter correction strategy was designed. Then, the control constraints were considered to form a novel T-MPSP algorithm. By combining the parameter correction strategy with the improved T-MPSP algorithm, an adaptive tracking guidance scheme was presented. Finally, simulations were carried out to show the effectiveness of the proposed method. In related future research, the saturation of the input should be considered in the design of the trajectory-tracking method.

Author Contributions: Conceptualization, C.O. and Z.C.; Methodology, C.O., Z.C. and Y.L.; Investigation, Z.C. and Y.L.; Data curation, C.S.; Writing—original draft, Z C.; Supervision, Y.L. All authors have read and agreed to the published version of the manuscript.

Funding: This study was co-supported in part by the National Natural Science Foundation of China (No. 61903146).

Data Availability Statement: Not applicable.

Acknowledgments: This study has been co-supported in part by the National Natural Science Foundation of China (No. 61903146 and 61873319).

Conflicts of Interest: The authors declare no conflict of interest.

Nomenclature

Scalar

h	altitude	v	velocity
γ	flight path angle	α	angle of attack (AOA)
m	mass	g	gravitational acceleration
I_{sp}	the specific impulse of the engine	T	the engine thrust
L	the aerodynamic lift	D	the aerodynamic drag
C_T	the thrust coefficient	C_L	the lift coefficient
C_D	the drag coefficient	ρ	the atmospheric density
ϕ	the throttle	S_{ref}	the reference area
Ma	mach number	ε	the terminal deviation thresholds
ω_a	uncorrelated white Gaussian noise	C_z	normal force coefficients
C_x	axial force coefficients	k	the current time instant

Matrix

F	the Jacobian matrix	P	the error covariance matrix
Φ	the state transition matrix	Q_k	the noise covariance matrix
K	the Kalman filter gain coefficient matrix	R_k	the noise covariance matrix

Subscripts

max	the maximum value	min	the minimum value
f	the final value and the superscript	$*$	the desired value
a	the new augmented state		

References

1. Chai, R.; Tsourdos, A.; Savvaris, A.; Chai, S.; Chen, C. Review of advanced guidance and control algorithms for space/aerospace vehicles. *Prog. Aerosp. Sci.* **2021**, *122*, 100696. [CrossRef]
2. Gharib, M.R.; Koochi, A.; Ghorbani, M. Path tracking control of electromechanical micro-positioner by considering control effort of the system. *Proc. Inst. Mech. Eng. Part I J. Syst. Control Eng.* **2021**, *235*, 984–991. [CrossRef]
3. Salehi Kolahi, M.R.; Gharib, M.R.; Heydari, A. Design of a non-singular fast terminal sliding mode control for second-order nonlinear systems with compound disturbance. *Proc. Inst. Mech. Eng. Part C J. Mech. Eng. Sci.* **2021**, *235*, 7343–7352. [CrossRef]
4. Abdalla, T.Y.; Abdulkarem, A. PSO-based optimum design of PID controller for mobile robot trajectory tracking. *Int. J. Comput. Appl.* **2012**, *47*, 30–35. [CrossRef]
5. Madhushani, T.; Maithripala, D.S.; Berg, J.M. Feedback regularization and geometric PID control for trajectory tracking of mechanical systems: Hoop robots on an inclined plane. In Proceedings of the 2017 American Control Conference (ACC), Seattle, WA, USA, 24–26 May 2017; pp. 3938–3943.
6. Rabah, M.; Rohan, A.; Han, Y.-J.; Kim, S.-H. Design of fuzzy-PID controller for quadcopter trajectory-tracking. *Int. J. Fuzzy Log. Intell. Syst.* **2018**, *18*, 204–213. [CrossRef]
7. Dang, T.S.; Duong, D.T.; Le, V.C.; Banerjee, S. A combined backstepping and adaptive fuzzy PID approach for trajectory tracking of autonomous mobile robots. *J. Braz. Soc. Mech. Sci. Eng.* **2021**, *43*, 156.
8. Shen, G.; Xia, Y.; Ma, D.; Zhang, J. Adaptive sliding-mode control for Mars entry trajectory tracking with finite-time convergence. *Int. J. Robust Nonlinear Control* **2019**, *29*, 1249–1264. [CrossRef]
9. Ai, X.; Yu, J. Fixed-time trajectory tracking for a quadrotor with external disturbances: A flatness-based sliding mode control approach. *Aerosp. Sci. Technol.* **2019**, *89*, 58–76. [CrossRef]
10. Sun, L.; Liu, Y. Fixed-time adaptive sliding mode trajectory tracking control of uncertain mechanical systems. *Asian J. Control* **2020**, *22*, 2080–2089. [CrossRef]
11. Zhou, B.; Huang, B.; Su, Y.; Zheng, Y.; Zheng, S. Fixed-time neural network trajectory tracking control for underactuated surface vessels. *Ocean Eng.* **2021**, *236*, 109416. [CrossRef]
12. Shen, C.; Shi, Y.; Buckham, B. Nonlinear model predictive control for trajectory tracking of an AUV: A distributed implementation. In Proceedings of the 2016 IEEE 55th Conference on Decision and Control (CDC), Las Vegas, NV, USA, 12–14 December 2016; pp. 5998–6003.
13. Nascimento, T.P.; Dórea, C.E.T.; Gonçalves, L.M.G. Nonlinear model predictive control for trajectory tracking of nonholonomic mobile robots: A modified approach. *Int. J. Adv. Robot. Syst.* **2018**, *15*, 1729881418760461. [CrossRef]
14. Chai, J.; Medagoda, E.; Kayacan, E. Adaptive and Efficient Model Predictive Control for Booster Reentry. *J. Guid. Control Dyn.* **2020**, *43*, 2372–2382. [CrossRef]
15. Emami, S.A.; Banazadeh, A. Fault-tolerant predictive trajectory tracking of an air vehicle based on acceleration control. *IET Control Theory Appl.* **2019**, *14*, 750–762. [CrossRef]

25. Padhi, R.; Kothari, M. Model predictive static programming: A computationally efficient technique for suboptimal control design. *Int. J. Innov. Comput. Inf. Control* **2009**, *5*, 399–411.
26. Powell, W.B. *Approximate Dynamic Programming: Solving the Curses of Dimensionality*; John Wiley & Sons: Hoboken, NJ, USA, 2007; Volume 703.
27. Halbe, O.; Raja, R.G.; Padhi, R. Robust reentry guidance of a reusable launch vehicle using model predictive static programming. *J. Guid. Control Dyn.* **2014**, *37*, 134–148. [CrossRef]
28. Zheng, H.; Hong, H.; Tang, S. Model predictive static programming rendezvous trajectory generation of unmanned aerial vehicles. In Proceedings of the 2019 IEEE International Conference on Unmanned Systems (ICUS), Beijing, China, 17–19 October 2019; pp. 415–420.
29. Wang, Y.; Hong, H.; Tang, S. Geometric control with model predictive static programming on SO (3). *Acta Astronaut.* **2019**, *159*, 471–479. [CrossRef]
30. Tripathi, A.K.; Padhi, R. Autonomous landing for uavs using t-mpsp guidance and dynamic inversion autopilot. *IFAC-PapersOnLine* **2016**, *49*, 18–23. [CrossRef]
31. Wang, M.; Zhang, S. Adaptive trajectory tracking algorithm based on tracking model-predictive-static-programming. *Acta Aeronaut. Astronaut. Sin.* **2018**, *39*, 322105–322113.
32. Kara, T.; Mary, A. Robust trajectory tracking control of robotic manipulators based on model-free PID-SMC approach. *J. Eng. Res.* **2018**, *6*, 170–188.
33. Emami, S.A.; Banazadeh, A. Intelligent trajectory tracking of an aircraft in the presence of internal and external disturbances. *Int. J. Robust Nonlinear Control.* **2019**, *29*, 5820–5844. [CrossRef]
34. Dou, L.; Su, X.; Zhao, X.; Zong, Q.; He, L. Robust tracking control of quadrotor via on-policy adaptive dynamic programming. *Int. J. Robust Nonlinear Control.* **2021**, *31*, 2509–2525. [CrossRef]
35. Qiu, X.; Hua, C.; Chen, J.; Zhang, L.; Guan, X. Model-free adaptive iterative sliding mode control for a robotic exoskeleton trajectory tracking system. *Int. J. Syst. Sci.* **2020**, *51*, 1782–1797. [CrossRef]
36. Murillo, O.J. A Fast Ascent Trajectory Optimization Method for Hypersonic Air-Breathing Vehicles. Ph.D. Thesis, Iowa State University, Ames, IA, USA, 2010.
37. Lopez-Sanchez, I.; Montoya-Chairez, J.; Perez-Alcocer, R.; Moreno-Valenzuela, J. Experimental Parameter Identifications of a Quadrotor by Using an Optimized Trajectory. *IEEE Access* **2020**, *8*, 167355–167370. [CrossRef]

Disclaimer/Publisher's Note: The statements, opinions and data contained in all publications are solely those of the individual author(s) and contributor(s) and not of MDPI and/or the editor(s). MDPI and/or the editor(s) disclaim responsibility for any injury to people or property resulting from any ideas, methods, instructions or products referred to in the content.

Article

Real-Time Trajectory Planning for Hypersonic Entry Using Adaptive Non-Uniform Discretization and Convex Optimization

Jiarui Ma, Hongbo Chen *, Jinbo Wang and Qiliang Zhang

School of System Science and Engineering, Sun Yat-Sen University, Guangzhou 510006, China
* Correspondence: chenhongbo@mail.sysu.edu.cn

Abstract: This paper introduces an improved sequential convex programming algorithm using adaptive non-uniform discretization for the hypersonic entry problem. In order to ensure real-time performance, an inverse-free precise discretization based on first-order hold discretization is adopted to obtain a high-accuracy solution with fewer temporal nodes, which would lead to constraint violation between the temporal nodes due to the sparse time grid. To deal with this limitation, an adaptive non-uniform discretization is developed, which provides a search direction for purposeful clustering of discrete points by adding penalty terms in the problem construction process. Numerical results show that the proposed method has fast convergence with high accuracy while all the path constraints are satisfied over the time horizon, thus giving potential to real-time trajectory planning.

Keywords: hypersonic entry; real-time trajectory planning; sequential convex programming; adaptive non-uniform discretization; feasibility guarantee

MSC: 90C25; 49M25

1. Introduction

With the increasing demand for spacecraft autonomy, not only the accuracy and optimality of the result trajectory but also the stability and efficiency of the solving algorithm are of crucial importance for real-time guidance. For the hypersonic entry problem, trajectory optimization is a popular method due to its extensive application prospect, yet it is still very challenging because of the highly nonlinear and complicated dynamics and constraints involved [1–5].

For the trajectory optimization problem, the existing methods can be classified as indirect methods and direct methods [6]. Based on Pontryagin's maximum principle [7], indirect methods derive the necessary conditions of optimality and solve a two-point boundary value problem (TPBVP) to obtain the result trajectory [8,9]. Due to the inherent nonlinearity in dynamics of the hypersonic entry problem, the solution of TPBVP presents significant challenges and falls short of real-time requirements [10]. Despite theoretical guarantees on the optimality of the solution, practical considerations limit its feasibility. The direct methods convert the original problem into an approximated finite parameter optimization problem and use a nonlinear programming (NLP) algorithm to solve it [11–13]. Although NLP-based methods have been successful in many applications, the time-consuming solution process and the lack of convergence guarantee are major challenges [10]. Furthermore, an appropriate initial guess should be selected to achieve a high-quality solution, especially for the hypersonic entry problem.

In recent years, the application of convex optimization in trajectory optimization has shown great potential in many problem, including powered descent guidance [14] and spacecraft rendezvous and proximity operations [15]. If a problem can be formulated in a convex form, it can be solved in polynomial time with a strong convergence guarantee

while obtaining the global optimal solution [16,17]. There are many state-of-the-art solvers based on the interior-point method (IPM) [18], such as GUROBI [19], MOSEK [20], and ECOS [21], to solve convex optimization problems efficiently. However, the majority of the trajectory optimization problems are non-convex. To deal with the nonlinearity in dynamics and constraints, the sequential convex programming (SCP) technique is developed to solve a sequence of convex subproblems to approximate the original problem [22]. The subproblems are formulated by linearization about a reference trajectory and subsequently solved via iterative refinement of the reference trajectory until convergence of the solution is achieved. In order to ensure the convex approximation is accurate, a trust region is imposed to make sure that the solution is not far from the reference trajectory. There are both hard and soft methods to address the trust region [23,24]. Recently, this SCP technique has been applied to hypersonic re-entry problems and has shown its effectiveness to obtain high-accuracy solutions. Liu formulated the entry problem as a second-order cone programming (SOCP) problem and applies the successive convexification method to solve it [25]. Wang and Grant proposed an improved SCP algorithm and introduced a new control input for the entry problem [26,27]. Wang and Cui developed a rapid trajectory optimization algorithm with the pseudospectral method [28]. However, multiple hundreds of temporal nodes are usually required to maintain the accuracy due to the nonlinearity of the hypersonic entry problem and the long flight duration. Achieving a balance between real-time performance and accuracy represents a significant challenge, as reducing the number of discrete points may result in a loss of precision.

To deal with this issue, Kamath and Açıkmeşe et al. [29–31] propose an inverse-free precise discretization based on first-order hold (FOH) discretization. Considering the consistency of the original non-convex dynamics with the reference trajectory and addition of the *stitching condition*, this discretization would guarantee high accuracy with few temporal nodes and has been effectively applied to various problems, including powered descent guidance [29], multi-phase rocket landing [30], and hypersonic entry guidance [31]. In [31], the amount of the temporal nodes is only 40 to achieve the commensurate accuracy in [27], which necessitated an excess of 200 nodes. Nevertheless, for uniform discretization, a reduced number of temporal nodes would generate a sparse time grid, which may lead to constraint violation between the temporal nodes since the constraints are only imposed at discrete points in the SCP subproblems, resulting in a new issue. In [30], an non-uniform discretization with additional time interval dilation variables is introduced in multi-phase rocket landing to adaptively decide the turning points of different phases. A similar idea is extended to the hypersonic entry problem and the penalized trust region (PTR) algorithm, a soft trust region method, is used to construct the SCP process [31]. However, the resulting trajectory would still experience constraint violation. In addition, in our experiment for the hypersonic entry problem, both hard and soft trust region methods with additional time interval variables showed worse convergence compared with those with uniform discretization. One of the reasons is that the dynamics are time-sensitive, and the other is that an effective search direction should be given to achieve a purposeful distribution of discrete points.

In this paper, we propose a novel adaptive non-uniform discretization method to handle the above issues.

- An inverse-free precise discretization is adopted to obtain high accuracy with few temporal nodes for real-time performance.
- An adaptive non-uniform discretizaiton is proposed to construct the SCP subproblem with additional penalty terms. This would give the solver a search direction to cluster the temporal nodes more purposefully and the propagated trajectory would satisfy all the path constraints as a result, which is the main contribution of this paper.
- The validity of proposed method is substantiated through a numerical experiment compared with other SCP methods.

This paper is organized as follows. Section 2 presents the model of the hypersonic entry trajectory optimization problem including the dynamics and constraints. Section 3 introduces the adaptive non-uniform discretization and constructs the SCP subproblem and iteration process. Simulation and results analysis are shown in Section 4. The conclusions are summarized in Section 5.

2. Problem Formulation

In this section, we consider a typical entry trajectory problem for an unpowered hypersonic vehicle with multiple path constraints.

2.1. 3-DoF Entry Dynamics

The dimensionless dynamic equations over a spherical, rotating Earth can be modeled as follows. More details can be referred to in [27].

$$\dot{x} = f(x,u) = \begin{cases} \dot{r} = V \sin \gamma \\ \dot{\theta} = \dfrac{V \cos \gamma \sin \psi}{r \cos \phi} \\ \dot{\phi} = \dfrac{V \cos \gamma \cos \psi}{r} \\ \dot{V} = -D - \dfrac{L \sin \gamma}{r^2} + \Omega_V \\ \dot{\gamma} = \dfrac{L \cos \sigma}{V} + \dfrac{(V^2 - 1/r) \cos \gamma}{Vr} + \Omega_\gamma \\ \dot{\psi} = \dfrac{L \sin \sigma}{V \cos \gamma} + \dfrac{V \cos \gamma \sin \psi \tan \phi}{r} + \Omega_\psi \end{cases} \tag{1}$$

where the state vectors are $x = [r, \theta, \phi, V, \gamma, \psi]$, representing the orbital radius, longitude, latitude, relative velocity, flight path angle, and heading angle, respectively. The Earth rotation-dependent terms, $\Omega_V, \Omega_\gamma, \Omega_\psi$, and the lift and drag accelerations, L, D, in (1) are shown below.

$$\begin{aligned} \Omega_V &= \Omega^2 r \cos \phi (\sin \gamma \cos \phi - \cos \gamma \sin \phi \cos \psi) \\ \Omega_\gamma &= 2\Omega \cos \phi \sin \psi + \Omega^2 r \cos \phi (\cos \gamma \cos \phi + \sin \gamma \sin \phi \cos \psi)/V \\ \Omega_\psi &= -2\Omega(\tan \gamma \cos \psi \cos \phi - \sin \phi) + \Omega^2 r \sin \phi \cos \phi \sin \psi /(V \cos \gamma) \\ L &= R_0 \rho V^2 S_{ref} C_L /(2m) \\ D &= R_0 \rho V^2 S_{ref} C_D /(2m) \end{aligned} \tag{2}$$

where Ω is the Earth self-rotation rate, $\rho = \rho_0 e^{(-h/hs)}$ is the atmospheric density depending on the altitude h, S_{ref}, m is the reference area and mass of the vehicle, and C_L, C_D are the aerodynamic lift and drag coefficients related to Mach number and the attack angle α.

As in [27], the control variable is restricted to bank angle $u = \sigma$. The attack angle α is pre-specified as a function of Mach number, as described in Section 4. All the variables are dimensionless and the dimensionless factors are shown in Table 1, where $R_0 = 6378.0$ km and $g_0 = 9.81$ m/s^2 represent the Earth's radius and the acceleration of gravity, respectively.

Table 1. The dimensionless factors' values.

Variable	Unit	Value
Time	s	$\sqrt{R_0/g_0}$
Distance	m	R_0
Velocity	m/s	$\sqrt{R_0 g_0}$
Acceleration	m/s^2	g_0
Angle	rad	1
Angle rate	rad/s	$\sqrt{g_0/R_0}$

2.2. State, Control, and Path Constraints

The inital and terminal conditions are

$$x(t_0) = x_0$$
$$x(t_f) = x_f \tag{3}$$

The control bounds and control rate constraints are given as follows:

$$-\sigma_{max} \leq \sigma \leq \sigma_{max}$$
$$-du_{max} \leq \dot{\sigma} \leq du_{max} \tag{4}$$

where σ_{max} and du_{max} are the bounds of the bank angle and its rate, respectively.

Three typical path constraints, including heat rate, dynamic pressure, and normal load, are considered:

$$\dot{Q} = p_1(r, V) = k_Q \sqrt{\rho} V^{3.15} \leq \dot{Q}_{max}$$
$$q = p_2(r, V) = 0.5\rho V^2 \leq q_{max} \tag{5}$$
$$n = p_3(r, V) = \sqrt{L^2 + D^2} \leq n_{max}$$

In this paper, no-fly zone constraints are considered as well, which are defined as cylinder zones with center $(\theta_{NFZ}, \phi_{NFZ})$, radius R_{NFZ}, and infinite altitude. Thus, the no-fly zone constraints are expressed as

$$(\theta - \theta_{NFZ})^2 + (\phi - \phi_{NFZ})^2 \geq R_{NFZ}^2 \tag{6}$$

2.3. Nonconvex Optimal Control Problem

The maximum terminal velocity hypersonic entry trajectory optimization problems with fixed flight time are considered in this paper, which is the same as [27]. The nonconvex optimal control problem is shown in Problem 1.

Problem 1.

$$\min_{x,u} J = -V(t_f) \tag{7}$$

s.t. (1), (3)–(6)

3. Improved SCP Method with Adaptive Non-Uniform Discretization

In this section, we introduce the improved SCP algorithm using adaptive non-uniform discretization for the hypersonic entry problem. In the interest of completeness, we provide a brief introduction to the non-uniform scheme and precise discretization technique, both of which, as in [31], are utilized in the proposed method. Further details will be presented subsequently. In order to seek an appropriate search direction and achieve a purposeful distribution of temporal nodes, additional penalty terms with respect to the nonlinear term of path constraints and the distance term from the trajectory to the no-fly zone center are considered in the SCP subproblem construction.

3.1. Time Interval Dilation

To introduce the non-uniform discretization, we consider the original nonlinear dynamics in the sub-interval $[t_k, t_{k+1})$, $k = 1, \ldots, N-1$,

$$\dot{x}(t) = f(t, x(t), u(t)), \quad t \in [t_k, t_{k+1}) \tag{8}$$

where $t_0 = t_1 < t_2 < \cdots < t_N = t_f$, and define an affine map to normalize the orginal time interval (may not be equal) to a fixed interval, $[0, 1)$:

$$\tau_k(t) = \frac{t - t_k}{s_k}, \quad t \in [t_k, t_{k+1}) \tag{9}$$

where $s_k = t_{k+1} - t_k$ is the length of the kth time interval and can be referred to as the time interval dilation [30]. So far, the dynamics Equation (8) can be rewritten with respect to the normalized time τ_k:

$$\dot{x}(\tau_k) = s_k f(\tau_k, x(\tau_k), u(\tau_k)) = F(\tau_k, x(\tau_k), u(\tau_k), s_k), \ \tau_k \in [0,1) \tag{10}$$

By treating s_k as additional decision variables, the solver is allowed to decide the adaptive time grids rather than a uniform temporal grid.

What is more, some exact constraints should be added in practical implementation to ensure physical meaning:

$$0 < \Delta_{\min} \leq s_k \leq \Delta_{\max} \tag{11}$$

to ensure the time order $t_0 = t_1 < t_2 < \cdots < t_N = t_f$ and adjacent temporal nodes are not far away, and

$$\sum_{k=1}^{N} s_k = t_f \tag{12}$$

to ensure the fixed flight time.

3.2. Convexification and Discretization

A convex approximation of the dynamic (10) can be obtained by the first-order Tylor expansion about a reference trajectory $(\bar{x}, \bar{u}, \bar{s})$. The approximate equation is a linear time-varing (LTV) system as follows:

$$\dot{x}(\tau_k) \approx A(\tau_k)x(\tau_k) + B(\tau_k)u(\tau_k) + S(\tau_k)s_k + d(\tau_k) \tag{13}$$

where $A(\tau_k), B(\tau_k), S(\tau_k)$ are the Jacobians of the dynamics with state control and time dilation, respectively.

$$\begin{aligned}
A(\tau_k) &\triangleq \nabla_x F(\tau_k, \bar{x}(\tau_k), \bar{u}(\tau_k), \bar{s}_k) \\
B(\tau_k) &\triangleq \nabla_u F(\tau_k, \bar{x}(\tau_k), \bar{u}(\tau_k), \bar{s}_k) \\
S(\tau_k) &\triangleq \nabla_{s_k} F(\tau_k, \bar{x}(\tau_k), \bar{u}(\tau_k), \bar{s}_k) \\
d(\tau_k) &\triangleq F(\tau_k, \bar{x}(\tau_k), \bar{u}(\tau_k), \bar{s}_k) \\
&\quad - A(\tau_k)\bar{x}(\tau_k) - B(\tau_k)\bar{u}(\tau_k) - S(\tau_k)\bar{s}_k
\end{aligned} \tag{14}$$

For discretization, a precise inverse-free discretization technique based on first-order hold (FOH) is adopted [29–31]. In the FOH case, the control input signal is considered as a piecewise affine function; thus, the control variables are only defined at the discrete time nodes and the control signal in the sub-interval can be parameterized as follows:

$$u(\tau_k) = (1 - \tau_k)u_k + \tau_k u_{k+1}, \ k = 1, \ldots, N-1 \tag{15}$$

where $t \in [t_k, t_{k+1})$ and $\tau_k \in [0,1)$ as given in (9). Thus, The LTV dynamics (13) can be easily rewritten with respect to the deviations from the reference trajectory.

$$\Delta\dot{x}(\tau_k) = A(\tau_k)\Delta x(\tau_k) + B(\tau_k)(1-\tau_k)\Delta u_k + B(\tau_k)\tau_k \Delta u_{k+1} + S(\tau_k)\Delta s_k \tag{16}$$

where $\bar{\square}$ denotes the reference quantity, $\Delta\square$ denotes the deviations from the reference trajectory, i.e., $\Delta\square = \square - \bar{\square}$ and $\Delta\dot{x}(\tau_k) = \dot{x}(\tau_k) - F(\tau_k, \bar{x}(\tau_k), \bar{u}(\tau_k), \bar{s}_k)$, and the coefficient matrixes A, B, S are the same as (14). It can be considered that the reference trajectory in the sub-interval $[t_k, t_{k+1})$ is in accordance with the original dynamics (10), rather than the convex approximation (13), like the typical FOH discretization in [32].

According to the knowledge of the linear system [33], the unique solution of (16) for $t \in [t_k, t_{k+1})$ and $\tau_k \in [0,1)$ is

$$\Delta x(\tau_k) = A_k(\tau_k)\Delta x(0) + B_k^-(\tau_k)\Delta u_k + B_k^+(\tau_k)\Delta u_{k+1} + S_k\Delta s_k \tag{17}$$

where
$$A_k(\tau_k) = \Phi(\tau_k, 0),$$
$$B_k^-(\tau_k) = A_k(\tau_k) \int_0^{\tau_k} \Phi^{-1}(\zeta, 0) B(\zeta)(1-\zeta) \, d\zeta,$$
$$B_k^+(\tau_k) = A_k(\tau_k) \int_0^{\tau_k} \Phi^{-1}(\zeta, 0) B(\zeta) \zeta \, d\zeta, \qquad (18)$$
$$S_k(\tau_k) = A_k(\tau_k) \int_0^{\tau_k} \Phi^{-1}(\zeta, 0) S(\zeta) \, d\zeta,$$

where $\Phi(\tau_k, 0)$ is called the state transition matrix (STM) with the following properties: $\Phi(0,0) = I$, $\dot{\Phi}(\tau_k, 0) = A(\tau_k) \Phi(\tau_k, 0)$, and $\Phi^{-1}(\tau, \eta) = \Phi(\eta, \tau)$.

In order to eliminate the inversion operation to avoid numerical problems, the $B_k^{\mp}$, S_k in (18) have the closer forms, as shown in Thereom 1, which is not proven in [29–31].

Theorem 1. *The coefficient matrixes* (18) *of the linear time-varying system* (17) *have the inverse-free form:* $\forall \tau_k \in [0, 1)$:
$$B_k^-(\tau_k) = \int_0^{\tau_k} A(\zeta) B_k^-(\zeta) + (1-\zeta) B(\zeta) \, d\zeta,$$
$$B_k^+(\tau_k) = \int_0^{\tau_k} A(\zeta) B_k^-(\zeta) + \zeta B(\zeta) \, d\zeta, \qquad (19)$$
$$S_k(\tau_k) = \int_0^{\tau_k} A(\zeta) S_k(\zeta) + S(\zeta) \, d\zeta$$

Proof. Choosing the B_k^- as an example, then taking the derivative and invoking the chain rule yields
$$\begin{aligned}\frac{d}{d\tau_k} B_k^-(\tau_k) &= \frac{d}{d\tau_k} A_k(\tau_k) \int_0^{\tau_k} \Phi^{-1}(\zeta, 0) B(\zeta)(1-\zeta) \, d\zeta \\ &\quad + A_k(\tau_k) \Phi^{-1}(\tau_k, 0) B(\tau_k)(1-\tau_k) \\ &= A(\tau_k) \Phi(\tau_k, 0) \int_0^{\tau_k} \Phi^{-1}(\zeta, 0) B(\zeta)(1-\zeta) \, d\zeta \\ &\quad + I \cdot B(\tau_k)(1-\tau_k) \\ &= A(\tau_k) B_k^-(\tau_k) + B(\tau_k)(1-\tau_k)\end{aligned} \qquad (20)$$

The first and second equal signs come from the properties of STM, while the last one is a simplification of the original form of B_k^- from (18). With $B_k^-(0) = 0$, the inverse-free form of $B_k^-(\tau_k)$ is obtained as shown in (19). B_k^+ and S_k can be obtained by the same process. □

For simplicity, we define 0_k and 1_k as 0 and 1, respectively, which denote that t is in the sub-interval $[t_k, t_{k+1})$. Then evaluating the LTV system (13) at $\tau_k = 1_k^-$, we obtain
$$\Delta x(1_k^-) = A_k \Delta x(0_k) + B_k^- \Delta u_k + B_k^+ \Delta u_{k+1} + S_k \Delta s_k \qquad (21)$$

Since the reference trajectory may not satisfy the original dynamics (1) in the sub-interval $[t_k, t_{k+1})$, Equation (22) would give $N - 1$ trajectory segments, which makes a discontinuity occur in the temporal time nodes $t_2, \ldots, t_N$. The *stitching condition* is introduced to obtain a continuous trajectory over the time horizon, as shown in Figure 1.
$$\Delta x(1_k^-) + \bar{x}(1_k^-) = \Delta x(1_k) + \bar{x}(1_k) \qquad (22)$$

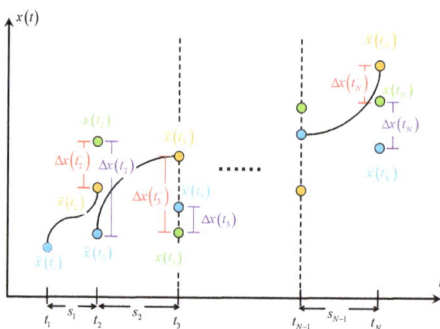

Figure 1. The discontinuity in the temporal nodes and the stitching condition, as in [31].

As a result, the discretized dynamics in terms of the deviations are as follows:

$$\Delta x_{k+1} = A_k \Delta x_k + B_k^- \Delta u_k + B_k^+ \Delta u_{k+1} + S_k \Delta s_k + x_{k+1}^{\text{prop}} - \bar{x}_{k+1} \quad (23)$$

where $\Delta x_k = \Delta x(0_k)$, $\bar{x}_{k+1} = \Delta x(1_k)$ and $x_{k+1}^{\text{prop}} = \bar{x}(1_k^-) = \bar{x}_k + \int_{0_k}^{1_k^-} F(\zeta, \bar{x}(\zeta), \bar{u}(\zeta), \bar{s}) \, d\zeta$, which denotes the integration of the original dynamic in $[t_k, t_{k+1})$ from the reference trajectory.

Thus the discretized dynamics with respect to absolute variables are recovered from Equation (23):

$$x_{k+1} = A_k x_k + B_k^- u_k + B_k^+ u_{k+1} + S_k s_k + x_{k+1}^{\text{prop}} - (A_k \bar{x}_k + B_k^- \bar{u}_k + B_k^+ \bar{u}_{k+1} + S_k \bar{s}_k) \quad (24)$$

With the idea of consistency of the original dynamics of the reference trajectory in sub-intervals and the addition of the *stitching condition*, the accuracy of the solution would be improved even over a sparse time grid.

3.3. Additional Penalty Terms in SCP Subproblem

In our experiments for the non-uniform scheme, it is observed that regardless of a hard trust region method with additional fixed constraints or a soft trust region method by augmenting the objective function with penalty terms, the method may not converge or converge very slowly, while the resulting propagated trajectory may violate the path constraints between temporal nodes as well.

As mentioned above, one reason is that an effective search direction should be given. Inspired by the PTR algorithm [24], we augment the objective function with additional penalty terms with respect to the nonlinear term of path constraints and the distance term of the no-fly zone, which would give a more purposeful direction to distribute the temporal nodes. As the path constraints exhibit high levels of nonlinearity, the logarithm transformation is used to mitigate this issue.

3.3.1. Log-Tranforms of Path Constraints

Consider the typical path constraints (5) of the hypersonic entry problem: heat rate, dynamic pressure, and normal load. Since the Tylor expansion of the original path constraints (5) would obtain complicated nonlinear terms, we take the logarithm transformation of both sides of (5):

$$\ln(\dot{Q}) = \ln(k_Q) + 0.5\ln(\rho_0) - \frac{R_0}{2h_s}(r-1) + 3.15\ln(V)$$
$$\leq \ln(\dot{Q}_{max})$$
$$\ln(q) = \ln(0.5) + \ln(\rho_0) - \frac{R_0}{h_s}(r-1) + 2\ln(V) \quad (25)$$
$$\leq \ln(q_{max})$$
$$\ln(n) = 0.5\ln(C_L^2 + C_D^2) + \ln(\frac{R_0 \rho_0 S_{ref}}{2m}) - \frac{R_0}{h_s}(r-1) + 2\ln(V)$$
$$\leq \ln(n_{max})$$

Due to the monotonicity of logarithmic transformations, Equation (25) is equivalent to (5), while the transformed constraints are linear to orbital radius r and only nonlinear to velocity V. Thus, the Tylor expansion of the transformed constraints has a simpler nonlinear term $\ln(V)$ than that of the original constraints.

Consider the second-order Tylor series expansion of $\ln(V)$:

$$\ln(V) = \ln(\overline{V}) + \frac{1}{\overline{V}}(V - \overline{V}) - \frac{1}{2\overline{V}^2}(V - \overline{V})^2 \quad (26)$$

Then we replace $\ln(V)$ in (25) with the primary term $\ln(\overline{V}) + \frac{1}{\overline{V}}(V - \overline{V})$ as the linearization of the transformed constraints and augment the objective function with the nonlinear term $\frac{1}{\overline{V}^2}(V - \overline{V})^2$, since the quadratic term is convex. With the variable time-step scheme and additional nonlinear penalty term, this would give the optimizer a target or direction to incentivize the temporal nodes to cluster around the highly nonlinear region.

3.3.2. Distance Penalty Term of No-Fly Zone

The propagated trajectory between temporal nodes could be within the no-fly zone because of the sparse time interval, as shown in Figure 2. In order to address the above issue, prior work is to set a dense time grid around the no-fly zone in advance. Instead of that, our method can allow discrete points to adaptively cluster around the no-fly zone during iteration, which would take full advantage of the non-uniform scheme.

The nonlinear no-fly zone constraints (6) can be linearized with first-order Tylor expansion:

$$2(\overline{\theta} - \theta_{NFZ})\theta + 2(\overline{\phi} - \phi_{NFZ})\phi \geq d \quad (27)$$

where $d = R_{NFZ}^2 - (\overline{\theta} - \theta_{NFZ})^2 - (\overline{\phi} - \phi_{NFZ})^2 + 2(\overline{\theta} - \theta_{NFZ})\overline{\theta} + 2(\overline{\phi} - \phi_{NFZ})\overline{\phi}$.

We penalize the distance from the trajectory to the center of the no-fly zone $\sum(\theta - \theta_{NFZ})^2 + (\phi - \phi_{NFZ})^2$ in the objective function, which would give the solver a search direction with physical significance. With the linear no-fly zone constraints (27), the discrete points would tend to cluster around the no-fly zone and disperse beyond it.

Note that the penalty term of the no-fly zone should be restricted to several temporal nodes to prevent all nodes from gathering around the no-fly zone.

3.4. Discrete Convex Subproblem and Iteration Algorithm

After the appeal discussion, we summarize the discrete convex sub-problem as shown in Problem 2. The control difference is regarded as the control rate constraint because of the FOH approximation of control. A hard trust region on time dilation is enforced in constraints, while the objective function is augmented with trust region terms of state and control, and two additional penalty terms: the nonlinear term of path constraints and distance penalty term of the no-fly zone. The trust term of time dilation can be omitted since the penalty terms of path constraints and the no-fly zone have shown a good iterative performance in our experiment. Thus, the SCP iteration process is given in Algorithm 1 as follows:

Algorithm 1: Improved SCP algorithm with adaptive non-uniform discretization

Input: Initial guess $\bar{x}, \bar{u}, \bar{s}$, convergence condition ϵ_x, ϵ_s, maximum iteration number $k_{\max}$

1. Set $k = 1$;
2. **while** $k \leq k_{\max}$ **do**
3.   Call for IPM solver to solve the subproblem **Problem 2** and obtain the solution $\hat{x}^k, \hat{u}^k, \hat{s}^k$.
4.   **if** $|\Delta_x| \leq \epsilon_x$ and $|\Delta_s| \leq \epsilon_s$ **then**
5.   Optimal trajectory $x^{opt} = \hat{x}^k, u^{opt} = \hat{u}^k, s^{opt} = \hat{s}^k$,
6.   **break**;
7.   **end**
8.   Update the reference trajectory $\bar{x} = x^k, \bar{u} = u^k, \bar{s} = s^k$.
9.   Set $k = k + 1$;
10. **end**

Result: Obtain optimal trajectory or reach maximum iteration.

Problem 2. *Discrete convex subproblem in SCP iteration*

$$\min_{x,u} J = -V_N + \underbrace{\omega_{tr,1} \sum_{k=1}^{N} \|x_k - \bar{x}_k\|_2^2 + \|u_k - \bar{u}_k\|_2^2}_{J_{tr,1}} + \underbrace{(\omega_{tr,2} \sum_{k=1}^{N-1} \|s_k - \bar{s}_k\|^2)}_{J_{tr,2}}$$

$$+ \underbrace{\omega_{nl} \sum_{k=1}^{N} \frac{\|V_k - \overline{V}_k\|^2}{\overline{V}_k^2}}_{J_{nl}} + \underbrace{\omega_{NFZ} \sum_{k=i_j^1}^{i_j^n} \|\theta_k - \theta_{NFZ}\|^2| + \|\phi_k - \phi_{NFZ}\|^2}_{J_{NFZ}}$$

s.t. $\forall k = 1, \ldots, N$

$x_{k+1} = A_k x_k + B_k^- u_k + B_k^+ u_{k+1} + S_k s_k + d_k$

$x_1 = x_0$

$x_N = x_f$

$-\sigma_{\max} \leq u_k \leq \sigma_{\max}$

$-du_{\max} \leq \dfrac{u_{k+1} - u_k}{\bar{s}_k} \leq du_{\max}$ (28)

$\bar{p}_{1,k} - \dfrac{R_0}{2h_s} r + \dfrac{3.15}{\overline{V}_k}(V - \overline{V}_k) \leq \ln(\dot{Q}_{\max})$

$\bar{p}_{2,k} - \dfrac{R_0}{h_s} r + \dfrac{2}{\overline{V}_k}(V - \overline{V}_k) \leq \ln(q_{\max})$

$\bar{p}_{3,k} - \dfrac{R_0}{h_s} r + \dfrac{2}{\overline{V}_k}(V - \overline{V}_k) \leq \ln(n_{\max})$

$2(\bar{\theta}_k - \theta_{NFZ})\theta_k + 2(\bar{\phi}_k - \phi_{NFZ})\phi_k \geq d_{NFZ}$

$\Delta_{\min} \leq s_k \leq \Delta_{\max}$

$\sum_{k=1}^{k-1} s_k = t_f$

$-\Delta T_{\max} \leq s_k - \bar{s}_k \leq \Delta T_{\max}$

where $\bar{p}_{i,k}, i = 1, 2, 3$ and d_{NFZ} can be obtained from (25) and (27), and $i_j^1, \ldots i_j^n$ are the temporal nodes set to cluster around the no-fly zone. Note that the augmented objective with $J_{tr,1}$ and $J_{tr,2}$ is the general PTR algorithm for non-uniform scheme.

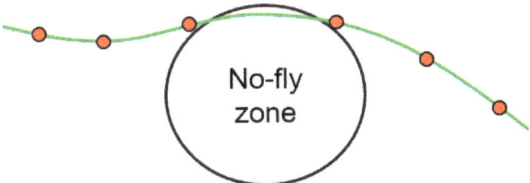

Figure 2. The propagated trajectory within the no-fly zone.

4. Numerical Results

In this section, the effectiveness of the proposed method is verified compared with a different SCP algorithm, as shown in Table 2. Two cases are considered to focus on distinct instances of constraint violation. Case 1 does not include the no-fly zone constraint and focuses on potential violations of path constraints as the number of discrete points decreases. With an additional no-fly zone constraint, Case 2 focuses on the phenomenon that the propagated trajectory may pass through the no-fly zone.

Table 2. Comparative SCP methods.

Name	Method	Reference
SCP1	non-uniform precise discretization	[31]
SCP2	uniform precise discretization	[29,32]
SCP3	uniform FOH discretization	[27]

The reference area and the mass of the vehicle are S_{ref} = 391.22 m^2 and m = 104,305.0 kg. The aerodynamic coefficients depend on the attack angle α (in degrees), while the angle-of-attack profile depends on the vehicle's velocity:

$$C_L = -0.041065 + 0.016292\,\alpha + 0.0002602\,\alpha^2$$
$$C_D = 0.080505 - 0.03026\,C_L + 0.86495\,C_L^2 \tag{29}$$

$$\alpha = \begin{cases} 40, & \text{if } V > 4570 \text{ m/s} \\ 40 - 0.20705(V - 4570)^2/340^2, & \text{else} \end{cases} \tag{30}$$

The remaining simulation parameters are shown in Table 3, which is the same as [27].

Table 3. Parameters for entry problem.

Parameter	Value	Parameter	Value
t_f	1600 s		
h_0	100 km	h_f	25 km
θ_0	0 deg	θ_f	12 deg
ϕ_0	0 deg	ϕ_f	70 deg
V_0	7450 m/s	γ_f	−10 deg
γ_0	0 deg	ψ_f	90 deg
ψ_0	0 deg	σ_{max}	80 deg
du_{max}	10 deg/s	k_Q	1.65 × 10^{-4}
$\dot{Q}_{max}$	1500 kW/m^2	q_{max}	18,000 N/m^2
n_{max}	2.5 g	Δ_{min}	10 s
Δ_{max}	150 s	ΔT_{max}	50 s
No-fly zone used in Case 2			
θ_{NFZ}	5 deg	ϕ_{NFZ}	50 deg
R_{NFZ}	5.5 deg		

The subproblem is constructed in YALMIP [34], a MATLAB modeling toolbox, and solved by ECOS [21], an open-source convex optimization solver. All of the numerical simulations are running on a personal desktop with an Intel Core i9 3.1 GHz processor.

The initial reference is the trajectory obtained by integrating the original dynamics (1) with the given initial control, as in [27]. The convergence condition is selected as $\Delta_x = \max_{1 \leq i \leq N} |\hat{x}_i^k - \bar{x}_i| \leq \epsilon_x = [1000 \text{ m}, 1 \text{ deg}, 1 \text{ deg}, 100 \text{ m/s}, 1 \text{ deg}, 1 \text{ deg}]$ and $\Delta_s = \max_{1 \leq i \leq N-1} |\hat{s}_i^k - \bar{s}_i| \leq \epsilon_s = 5 \text{ s}$, where $\hat{x}$ and $\hat{s}$ are the solution of the subproblem.

4.1. Iterative Performance

Comparisons of the state and control profiles for Case 1 and Case 2 are shown in Figures 3–8. The account of temporal nodes for the proposed method are as follows. SCP1 and SCP2 are each 40, in which case the constraint violation is observed, and SCP3 is 300 in order to maintain the same accuracy as the above methods.

It can be seen that the solutions of the proposed method are similar to those from SCP2 and SCP3. Note that the solutions of SCP1 are quite different due to its poor convergence. As shown in Figures 5 and 8, one shortcoming of the proposed method is that the control jitter in the segments where points cluster is obvious because of the dense time grid around those points.

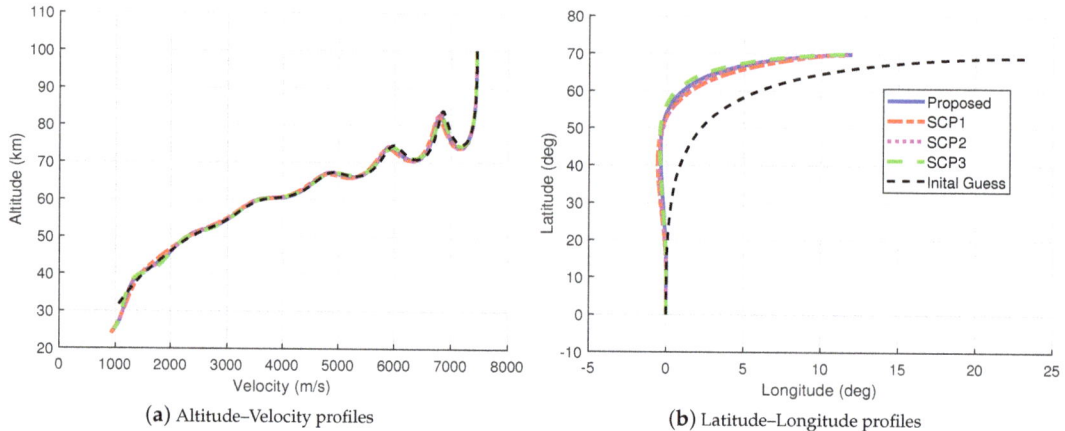

Figure 3. Comparisons of the altitude–velocity and latitude–longitude profiles for Case 1.

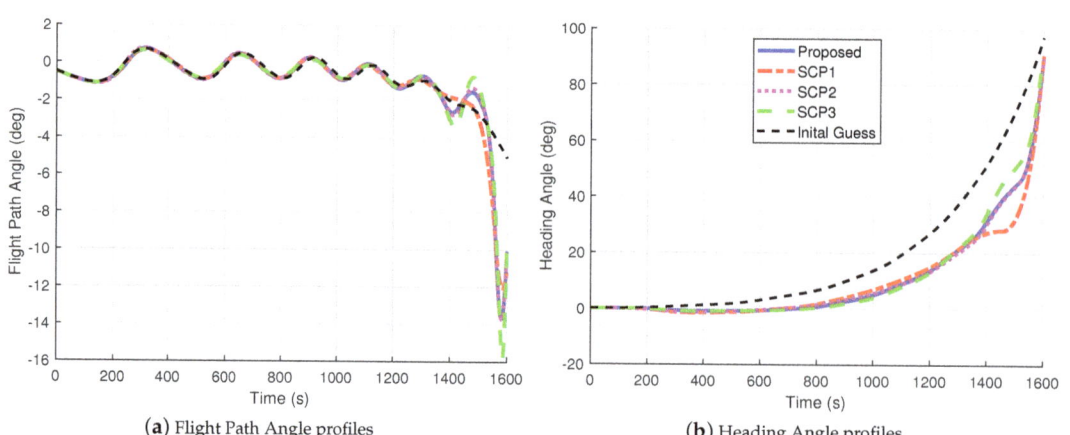

Figure 4. Comparisons of the flight path angle and heading angle profiles for Case 1.

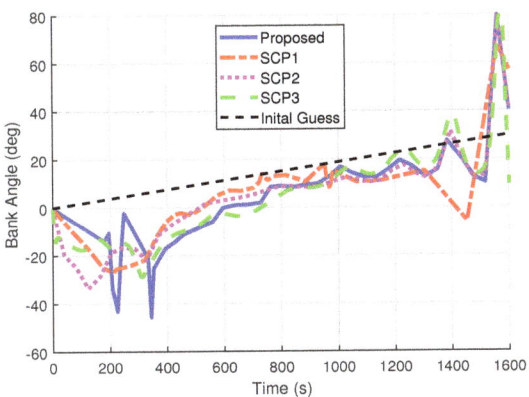

Figure 5. Comparisons of the control profile for Case 1.

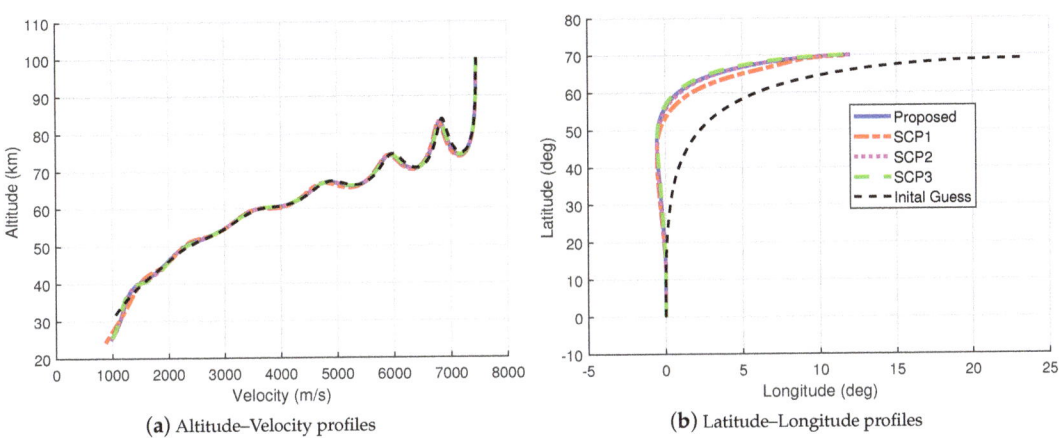

Figure 6. Comparisons of the altitude–velocity and latitude-longitude profiles for Case 2.

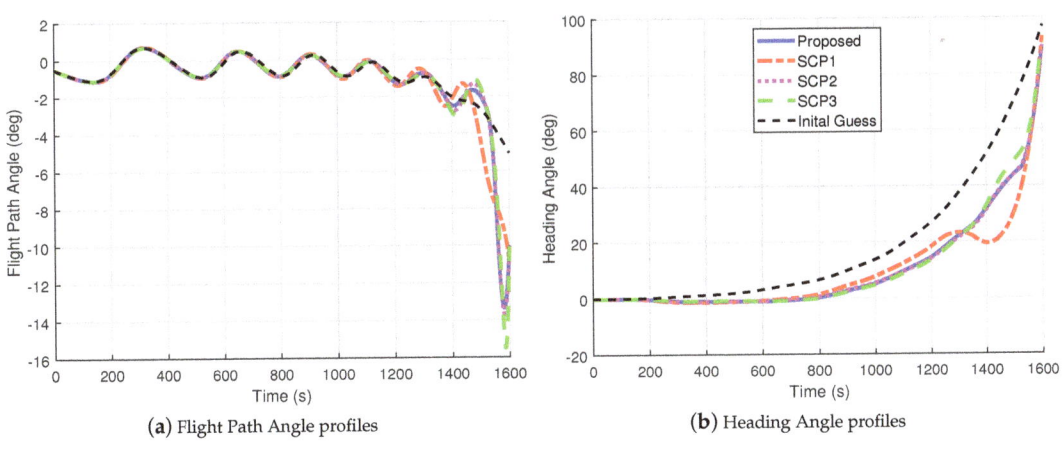

Figure 7. Comparisons of the flight path angle and heading angle profiles for Case 2.

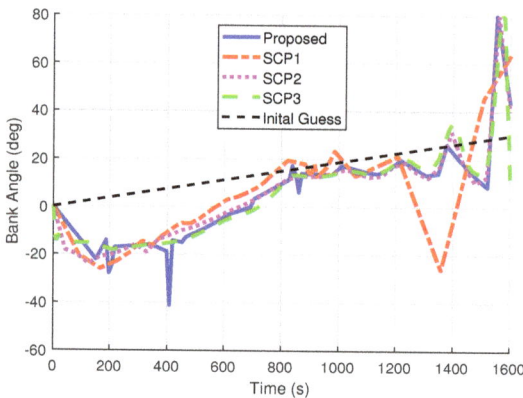

Figure 8. Comparisons of the control profile for Case 2.

The iterative performance of the proposed method and comparative methods is shown in Figure 9. The iteration number of the proposed method is less than the comparative methods. Note that the comparative methods would require more iterations to meet the convergence condition when the objective function is near the optimal value, while the proposed method required fewer iterations, demonstrating its fast convergence. What is more, for the non-uniform scheme, SCP1 with light penalty weighting $\omega_{tr,2} < 1$ in $J_{tr,2}$ reached the maximum number of iterations, while that with heavy weighting $\omega_{tr,2} > 1$ would show negligible alterations for the distribution of temporal nodes; thus, the addition of time interval dilation is not necessary in this case.

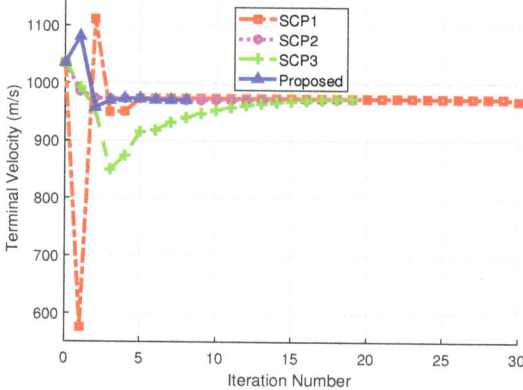

Figure 9. The terminal velocity in each iteration.

For the non-uniform scheme, the discrete point distribution and the change of time dilation, $\max |s - \bar{s}|$, with iterations are shown in Figure 10. It can be observed that there is no obvious clustering rule for SCP1, while the position of discrete points always changes with the number of iterations. In contrast, the discrete points of the proposed method would cluster after a few iterations. In addition, the results of time interval change with iterations show that the proposed method has stable convergence performance, since the change between two adjacent iterations decreases progressively, which means the result trajectory would become increasingly similar.

In order to guarantee that the result trajectory of the SCP process is feasible to meet the original dynamics, the residual error between the optimized results and the trajectory obtained by integrating the original dynamics is measured. The residual error results for Case 1 and Case 2 are shown in Tables 4 and 5, averaged over 50 simulation runs.

Table 4. Residual error with different temporal nodes for Case 1.

Method	Temporal Node	Iteration	CPU Time (s)	Δr (m)	$\Delta \theta$ (deg)	$\Delta \phi$ (deg)	ΔV (m/s)	$\Delta \gamma$ (deg)	$\Delta \psi$ (deg)
Proposed	40	8	1.409	22.577	0.044	0.011	1.910	0.022	0.072
	50	8	1.452	18.457	0.042	0.010	1.509	0.018	0.046
	60	10	1.758	3.798	0.035	0.005	1.249	0.019	0.027
SCP1	40	30	5.247	814.272	0.323	0.128	25.132	1.196	0.203
	50	30	5.790	627.624	0.218	0.105	28.145	0.288	1.544
	60	30	6.293	587.483	0.203	0.097	26.329	0.279	1.416
SCP2	40	14	2.085	5.201	0.005	0.002	0.236	0.001	0.010
	50	14	2.2370	21.356	0.011	0.006	0.851	0.004	0.055
	60	14	2.548	11.769	0.002	0.001	0.172	0.040	0.051
SCP3	200	15	5.5930	95.028	0.001	0.001	0.496	0.147	0.282
	300	19	9.857	42.689	0.004	0.002	0.048	0.034	0.085
	400	20	14.833	27.497	0.005	0.002	0.287	0.005	0.018

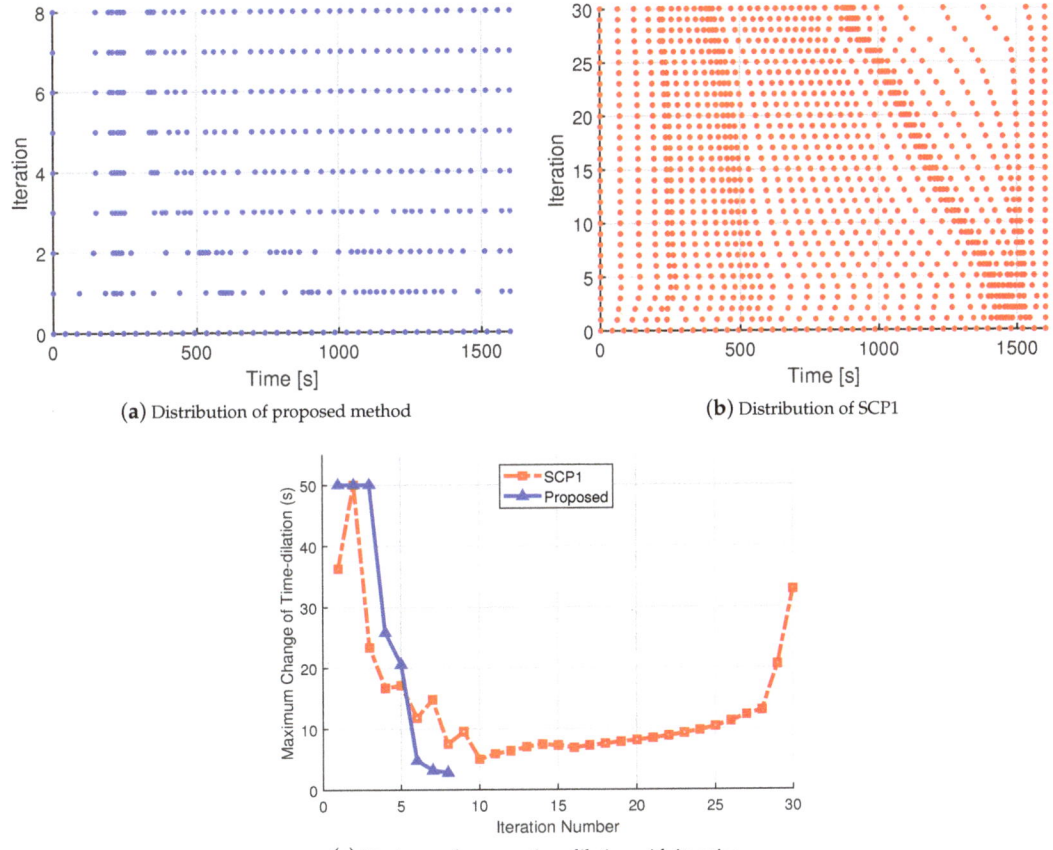

(**a**) Distribution of proposed method

(**b**) Distribution of SCP1

(**c**) Maximum change or time dilation with iteration

Figure 10. Discrete point distribution and time dilation change with iteration.

Table 5. Residual error for entry problem for Case 2.

Method	Temporal Node	Iteration	CPU Time (s)	Δr (m)	$\Delta\theta$ (deg)	$\Delta\phi$ (deg)	ΔV (m/s)	$\Delta\gamma$ (deg)	$\Delta\psi$ (deg)
Proposed	40	8	1.262	14.279	0.017	0.006	0.076	0.015	0.083
	50	8	1.403	9.201	0.025	0.007	0.426	0.022	0.032
	60	9	1.730	9.907	0.030	0.005	0.813	0.012	0.006
SCP1	40	30	4.234	1045.231	0.445	0.280	86.537	0.994	3.329
	50	30	4.532	654.915	0.192	0.111	29.263	0.279	1.672
	60	30	5.054	516.706	0.175	0.084	22.669	0.258	1.145
SCP2	40	11	1.437	5.142	0.002	0.004	0.234	0.002	0.015
	50	15	1.771	5.453	0.001	0.001	0.164	0.004	0.016
	60	15	1.932	5.285	0.001	0.0004	0.126	0.017	0.026
SCP3	200	11	3.891	97.585	0.002	0.001	0.682	0.224	0.573
	300	13	6.760	47.728	0.002	0.002	0.462	0.094	0.281
	400	15	10.334	36.177	0.005	0.006	0.586	0.057	0.182

It can be observed that for SCP2 and SCP3, the precise discretization can guarantee commensurate accuracy with fewer temporal nodes, while more than 200 nodes are needed to achieve the same result in [27], which demonstrates the effectiveness of the precise discretization. Note that, with the non-uniform scheme, SCP1 showed worse convergence performance and low accuracy as mentioned above, which means that the feasibility of the result trajectory is not guaranteed. In contrast, the proposed method overcomes the shortcomings, maintains the advantage of the precise discretization, and shows better convergence performance, as shown in Tables 4 and 5.

4.2. Constraint Satisfaction Performance

As mentioned above, for the uniform precise discretization, the propagated trajectory may violate the path constraints between temporal nodes due to the sparse time grid. This phenomenon was observed to occur when the account of temporal nodes decreased to 40. Thus, we choose the case of 40 nodes for presentation. The path constraints of the proposed method for Case 1 are shown in Figure 11 and contrasted with SCP1 and SCP2.

It can be observed that the heat load would touch the boundary during the initial state of flight. The constraint violations occur for the propagated trajectory of SCP1 and SPC2 due to the sparse time grid. In contrast, the propagated trajectory of the proposed method satisfies the path constraints over the time horizon. In addition, the discrete points would cluster around the peak value of heat load, as shown in Figure 11.

Furthermore, the same phenomenon may occur for the no-fly zone constraints as well. Case 2 focuses on the no-fly zone constraint violation, and the numerical results are shown in Figure 12. In this case, all the path constraints were satisfied for all methods in our experiments. It can be observed that the propagated trajectories of SCP1 and SCP2 both pass through the no-fly zone. Note that, for the non-uniform scheme, the discrete points of SCP1 do not cluster around the no-fly zone. The 3D trajectory of the proposed method is shown in Figure 12b; the points in red are set to cluster around the no-fly zone. The propagated trajectory of the proposed method skimmed over the no-fly zone, while the time interval between adjacent red points is the minimum set in (11), which indicated that the point distribution of the proposed method is better than SCP2.

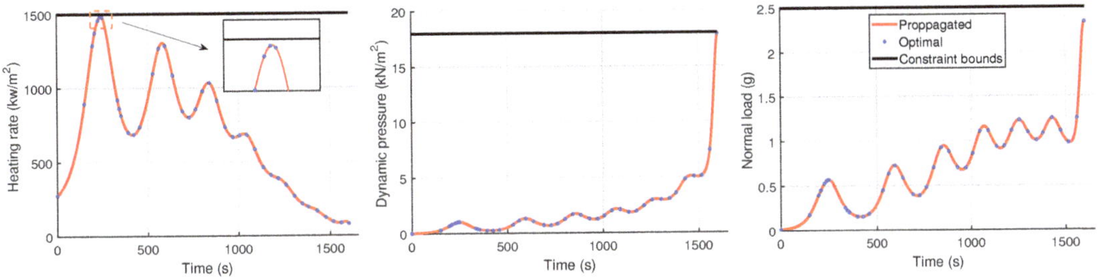

(a) Path Constraints for propagated trajectory of proposed method

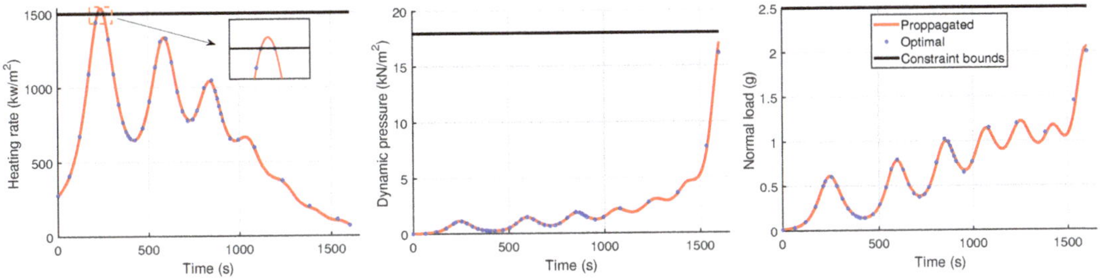

(b) Path Constraints for propagated trajectory of SCP1

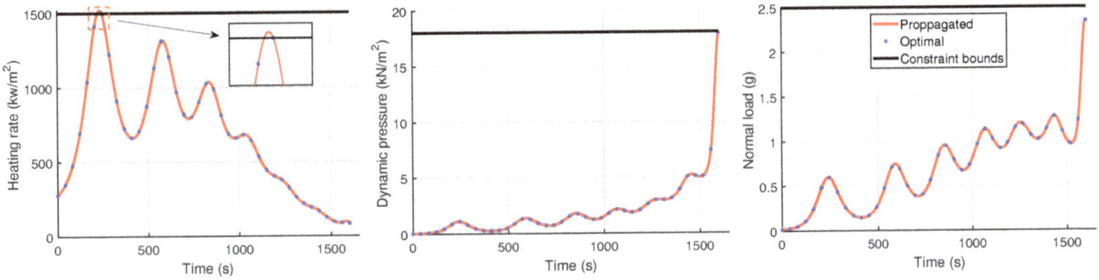

(c) Path Constraints for propagated trajectory of SCP2

Figure 11. Path constraints of propagated trajectories for Case 1.

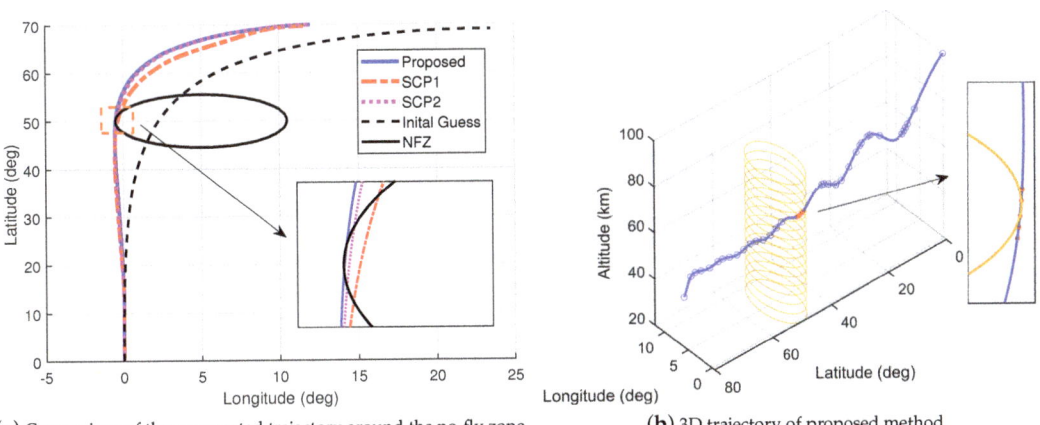

(a) Comparison of the propagated trajectory around the no-fly zone

(b) 3D trajectory of proposed method

Figure 12. The propagated trajectory for Case 2.

5. Conclusions

This paper proposes an improved SCP algorithm for the hypersonic entry problem using a novel adaptive non-uniform discretization. The proposed method has advantages in performance of path constraint satisfaction and convergence. Firstly, the proposed method employs an inverse-free precise discretization to ensure high accuracy and real time performance. Then, an adaptive non-uniform scheme is developed to distribute discrete points adaptively by adding additional penalty terms in the SCP subproblem, which would guarantee constraint satisfaction. Finally, numerical results show that the proposed method achieves a fast convergence while maintaining high accuracy with few temporal nodes. More importantly, the discrete points of the proposed method would cluster around the segment where the constraints may be violated, and the propagated trajectory satisfies all the path constraints over the time horizon even for a small number of discrete points.

Future work will focus on the following points: (1) Due to the non-uniform scheme, a similar idea can extend to the hypersonic entry problem with the waypoint constraint and other problems; (2) The simulation will be carried out on an embedded platform to verify the effectiveness of the proposed method for a limited-power environment; (3) High performance solvers are considered to further improve the solving speed, such as the proportional integral projected gradient method [35], a first-order method for the conic convex problem.

Author Contributions: Methodology, J.M.; Validation, J.W.; Writing—original draft preparation, J.M. Writing—review and editing, J.W., Q.Z. and H.C. All authors have read and agreed to the published version of the manuscript.

Funding: This research was funded by the Basic and Applied Basic Research Project of Guangzhou Municipal Science and Technology Bureau, grant number 202201011187.

Data Availability Statement: Not applicable.

Conflicts of Interest: The authors declare no conflict of interest.

References

1. Lu, P. Entry guidance and trajectory control for reusable launch vehicle. *J. Guid. Control Dyn.* **1997**, *20*, 143–149. [CrossRef]
2. Shen, Z.; Lu, P. Onboard generation of three-dimensional constrained entry trajectories. *J. Guid. Control Dyn.* **2003**, *26*, 111–121. [CrossRef]
3. Lu, P. Entry guidance: A unified method. *J. Guid. Control Dyn.* **2014**, *37*, 713–728. [CrossRef]
4. Jorris, T.R.; Cobb, R.G. Three-dimensional trajectory optimization satisfying waypoint and no-fly zone constraints. *J. Guid. Control Dyn.* **2009**, *32*, 551–572. [CrossRef]
5. Grant, M.J.; Braun, R.D. Rapid indirect trajectory optimization for conceptual design of hypersonic missions. *J. Spacecr. Rocket.* **2015**, *52*, 177–182. [CrossRef]
6. Betts, J.T. Survey of numerical methods for trajectory optimization. *J. Guid. Control Dyn.* **1998**, *21*, 193–207. [CrossRef]
7. Pontryagin, L.S. *Mathematical Theory of Optimal Processes*; CRC Press: Boca Raton, FL, USA, 1987.
8. Pan, B.; Lu, P.; Pan, X.; Ma, Y. Double-homotopy method for solving optimal control problems. *J. Guid. Control Dyn.* **2016**, *39*, 1706–1720. [CrossRef]
9. Zheng, Y.; Cui, H.; Ai, Y. Indirect trajectory optimization for mars entry with maximum terminal altitude. *J. Spacecr. Rocket.* **2017**, *54*, 1068–1080. [CrossRef]
10. Ben-Asher, J.Z. *Optimal Control Theory with Aerospace Applications*; American Institute of Aeronautics and Astronautics: Reston, VA, USA, 2010.
11. Fahroo, F.; Ross, I.M. Direct trajectory optimization by a Chebyshev pseudospectral method. *J. Guid. Control Dyn.* **2002**, *25*, 160–166. [CrossRef]
12. Kameswaran, S.; Biegler, L.T. Convergence rates for direct transcription of optimal control problems using collocation at Radau points. *Comput. Optim. Appl.* **2008**, *41*, 81–126. [CrossRef]
13. Garg, D.; Patterson, M.; Hager, W.W.; Rao, A.V.; Benson, D.A.; Huntington, G.T. A unified framework for the numerical solution of optimal control problems using pseudospectral methods. *Automatica* **2010**, *46*, 1843–1851. [CrossRef]
14. Acikmese, B.; Ploen, S.R. Convex programming approach to powered descent guidance for mars landing. *J. Guid. Control Dyn.* **2007**, *30*, 1353–1366. [CrossRef]

5. Lu, P.; Liu, X. Autonomous trajectory planning for rendezvous and proximity operations by conic optimization. *J. Guid. Control Dyn.* **2013**, *36*, 375–389. [CrossRef]
6. Boyd, S.P.; Vandenberghe, L. *Convex Optimization*; Cambridge University Press: Cambridge, UK, 2004.
7. Liu, X.; Lu, P.; Pan, B. Survey of convex optimization for aerospace applications. *Astrodynamics* **2017**, *1*, 23–40. [CrossRef]
8. Wright, S.J. *Primal-Dual Interior-Point Methods*; SIAM: Philadelphia, PA, USA, 1997.
9. Gurobi Optimization, Ltd. Gurobi Optimizer Reference Manual. 2021. Available online: https://www.gurobi.com/documentation/current/refman/index.html (accessed on 22 May 2023).
10. ApS, M. Mosek optimization toolbox for matlab. *User's Guide Ref. Man. Version* **2019**, *4*, 1.
11. Domahidi, A.; Chu, E.; Boyd, S. ECOS: An SOCP solver for embedded systems. In Proceedings of the 2013 European Control Conference (ECC), Zurich, Switzerland, 17–19 July 2013; pp. 3071–3076.
12. Mao, Y.; Szmuk, M.; Xu, X.; Açikmese, B. Successive convexification: A superlinearly convergent algorithm for non-convex optimal control problems. *arXiv* **2018**, arXiv:1804.06539.
13. Szmuk, M.; Acikmese, B.; Berning, A.W. Successive convexification for fuel-optimal powered landing with aerodynamic drag and non-convex constraints. In Proceedings of the AIAA Guidance, Navigation, and Control Conference, San Diego, CA, USA, 4–8 January 2016; p. 0378.
14. Malyuta, D.; Reynolds, T.P.; Szmuk, M.; Lew, T.; Bonalli, R.; Pavone, M.; Acikmese, B. Convex optimization for trajectory generation. *arXiv* **2021**, arXiv:2106.09125.
15. Liu, X.; Shen, Z.; Lu, P. Entry trajectory optimization by second-order cone programming. *J. Guid. Control Dyn.* **2016**, *39*, 227–241. [CrossRef]
16. Wang, Z.; Grant, M.J. Constrained trajectory optimization for planetary entry via sequential convex programming. *J. Guid. Control Dyn.* **2017**, *40*, 2603–2615. [CrossRef]
17. Wang, Z.; Lu, Y. Improved sequential convex programming algorithms for entry trajectory optimization. *J. Spacecr. Rocket.* **2020**, *57*, 1373–1386. [CrossRef]
18. Wang, J.; Cui, N.; Wei, C. Rapid trajectory optimization for hypersonic entry using a pseudospectral-convex algorithm. *Proc. Inst. Mech. Eng. Part G J. Aerosp. Eng.* **2019**, *233*, 5227–5238. [CrossRef]
19. Kamath, A.G.; Elango, P.; Kim, T.; Mceowen, S.; Yu, Y.; Carson, J.M.; Mesbahi, M.; Acikmese, B. Customized real-time first-order methods for onboard dual quaternion-based 6-DoF powered-descent guidance. In Proceedings of the AIAA SCITECH 2023 Forum, Orlando, FL, USA, 8–12 January 2023; p. 2003.
20. Kamath, A.G.; Elango, P.; Yu, Y.; Mceowen, S.; Carson, J.M., III; Açıkmeşe, B. Real-Time Sequential Conic Optimization for Multi-Phase Rocket Landing Guidance. *arXiv* **2022**, arXiv:2212.00375.
21. Mceowen, S.; Kamath, A.G.; Elango, P.; Kim, T.; Buckner, S.C.; Acikmese, B. High-Accuracy 3-DoF Hypersonic Reentry Guidance via Sequential Convex Programming. In Proceedings of the AIAA SCITECH 2023 Forum, San Diego, CA, USA, 8–12 January 2023; p. 0300.
22. Reynolds, T.P. *Computational Guidance and Control for Aerospace Systems*; University of Washington: Seattle, WA, USA, 2020.
23. Antsaklis, P.J.; Michel, A.N. *Linear Systems*; Springer: Berlin/Heidelberg, Germany, 1997; Volume 8.
24. Lofberg, J. YALMIP: A toolbox for modeling and optimization in MATLAB. In Proceedings of the 2004 IEEE International Conference on Robotics and Automation (IEEE Cat. No. 04CH37508), Taipei, Taiwan, 2–4 September 2004; pp. 284–289.
25. Yu, Y.; Elango, P.; Topcu, U.; Açıkmeşe, B. Proportional–integral projected gradient method for conic optimization. *Automatica* **2022**, *142*, 110359. [CrossRef]

Disclaimer/Publisher's Note: The statements, opinions and data contained in all publications are solely those of the individual author(s) and contributor(s) and not of MDPI and/or the editor(s). MDPI and/or the editor(s) disclaim responsibility for any injury to people or property resulting from any ideas, methods, instructions or products referred to in the content.

Article

Cubature Kalman Filters Model Predictive Static Programming Guidance Method with Impact Time and Angle Constraints Considering Modeling Errors

Zihan Xie, Jialun Pu *, Changzhu Wei and Yingzi Guan

School of Astronautics, Harbin Institute of Technology, Harbin 150000, China
* Correspondence: nosay@hit.edu.cn

Abstract: This paper proposes a CKF-MPSP guidance method for hitting stationary targets with impact time and angle constraints for missiles in the presence of modeling errors. This innovative guidance scheme is composed of three parts: First, the model predictive static programming (MPSP) algorithm is used to design a nominal guidance method that simultaneously satisfies impact time and angle constraints. Second, the cubature Kalman filter (CKF) is introduced to estimate values of the influence of the inevitable modeling errors. Finally, a one-step compensation scheme is proposed to eliminate the modeling errors' influence. The proposed method uses a real missile dynamics model, instead of a simplified one with a constant-velocity assumption, and eliminates the effects of modeling errors with the compensation scheme; thus, it is more practical. Simulations in the presence of modeling errors are conducted, and the results illustrate that the CKF-MPSP guidance method can reach the target with a high accuracy of impact time and angles, which demonstrates the high precision and strong robustness of the method.

Keywords: terminal guidance; impact time constraint; impact angle constraint; model predictive static programming; cubature Kalman filter; modeling errors

MSC: 37M10

1. Introduction

Guidance methods have always been a hot research topic in the field of missiles. The early guidance-law design considers only the minimized miss distance requirement, such as proportional navigation (PN). With the development of military science and technology, classical guidance methods no longer satisfy combat requirements [1]. There is an urge to research advanced guidance methods with multiple constraints. Impact angle and impact time constraints are essential for advanced terminal guidance methods. Impact angles are divided into path angle and azimuth angle. Attacking the target with a missile with proper impact angles may improve the destructive effect and hit weak parts of the target. The impact time is vital for attacking time-sensitive targets. Moreover, the guidance method with impact angle and time constraints gives the multi-missile cooperative guidance capability. Thus, investigating the impact time and angle-constrained guidance method (ITACG) is very important.

In this paper, a CKF-MPSP guidance method with impact time and angle constraints for a stationary target is proposed considering modeling errors. A baseline guidance method with impact time and angle constraints is designed based on the MPSP algorithm. The modeling errors are estimated by the CKF and compensate the baseline guidance method to eliminate their effects. The main contributions of this paper are shown below.

(1) An ITACG is designed for a stationary target based on the MPSP algorithm, which can simultaneously achieve impact time and angle constraints. This guidance method

considers the missile's dynamic model instead of a constant-velocity model. Therefore, the proposed method is more suitable for practical missiles.

(2) The proposed guidance method takes the desired time as a terminal condition for static planning. Time-to-go information is not required during the guidance process, which avoids the influence of time-to-go estimation errors on time control accuracy.

(3) A CKF-based modeling error compensation scheme is proposed to solve the problem of the MPSP algorithm being unable to be used for error conditions. This improvement enhances the feasibility of the guidance method in practical applications since the modeling error is inevitable.

It is worth noting that, although the CKF-MPSP guidance method is proposed in the scenario of a missile attacking a stationary target on the ground, in this paper, it can also be used for position control, such as aircraft landing. Furthermore, by predicting and introducing the target's motion model, the CKF-MPSP method can be used for moving-target interception. However, these applications are not the focus of this paper and require further research in the future.

This article is organized as follows. Section 2 provides a literature review of existing achievements in related research fields. Section 3 formulates the problem researched in this paper. Section 4 proposes the CKF-MPSP guidance method to implement time- and angle-constrained guidance. Simulation results are given in Section 5. Section 6 gives the conclusion.

2. Literature Review

Many scholars have conducted much research on ITACG problems. The mainstream methods can be divided into non-predictive guidance and predictive guidance methods.

The guidance laws that utilize the current relative motion information between missiles and targets are called non-predictive guidance. These kinds of methods only adopt the relative motion model instead of the actual model of the missile, making the design of such methods relatively simple. The existing non-predictive guidance methods basically follow two design paradigms.

The first design paradigm is to design guidance laws both in the line-of-sight (LOS) and normal LOS direction for time control and angle constraints, respectively. These kinds of methods are widely used in cooperative guidance scenarios. Zhang [2] proposed an ITACG with finite time convergence. Yu [1], Chen [3], and Lin [4] proposed fixed-time ITACGs based on the sliding mode theory. Ma [5] designed a disturbance-observer-based ITACG to enable the interception of maneuvering targets. Wang [6] proposed a decoupled three-dimensional sliding mode guidance law achieving simultaneous arrival at the target for multiple missiles with angle constraints. Jing [7] proposed a predefined-time convergence ITACG method for a multi-missile cooperative guidance scenario. Because of the design paradigm, the methods presented in Refs. [1–7] all need a control force both in the LOS and normal LOS direction for time control and angle constraints, respectively. However, most existing missiles are thrust-free and controlled by aerodynamic force in the terminal guidance period. In reality, missiles cannot provide the guidance command in the LOS direction, limiting the practical application of such guidance methods.

The second design paradigm is to design guidance laws only in the normal direction of LOS/velocity, which is more practical but more challenging compared to the above methods. Chen [8] simplified missile dynamics under a small heading error approximation and derived an optimal guidance law with impact time and angle constraints against a stationary target. Kim [9] introduced a polynomial guidance method considering impact time and angle constraints. Zhao [10] designed the trajectory as a function with two undetermined parameters and adjusted them to control the impact time and angle. Kang [11] derived a look-angle shaping scheme for ITACG. Hou [12,13] proposed a time-to-go estimation scheme for terminal sliding mode guidance with an impact angle constraint and further designed a nonsingular terminal sliding mode guidance law considering impact time and angle simultaneously. Chen [14] designed a two-stage guidance that satisfies time and angle constraints through a proper guidance-switching strategy. Zhang [15] proposed an ITACG by introducing an impact time feedback control term based on biased PN. Yan [16] proposed a computational geometry guidance against stationary targets satisfying the constraints by iterating parameters of the geometry curve. Majumder [17] proposed a sliding-mode-control-based nonlinear guidance scheme for controlling both impact angle and impact time simultaneously. Liu [18] designed an adaptive sliding mode ITACG method, increasing its adaptability and robustness. Wang [19] designed a two-stage guidance method, achieving ITACG through reasonable switching between two guidance rules. In Refs. [8–19], the guidance methods rely on time-to-go or range-to-go estimation. However, the estimation is difficult because of the uncontrollable varying velocity. To simplify the design process, the missile's velocity is assumed as constant, which may cause significant time estimation errors, especially when the trajectory is winding due to impact angle constraints. Thus, the guidance effect is not satisfactory in reality. Overall, there are difficulties in the practical application of the non-predictive ITACG methods in missiles because of the model mismatch.

Unlike non-predictive guidance, the predictive guidance methods predict the terminal states using a real model, which can avoid the model mismatch and thus can derive a better guidance performance. As one of the predictive guidance methods, the MPSP-based guidance methods have received widespread attention in recent years. The MPSP algorithm was first introduced in Ref [20]. Combining the philosophy of approximate dynamic programming and model predictive control, the MPSP algorithm obtains the terminal estimation of the output vector by integral prediction. Then, it efficiently solves the optimization problem by turning the dynamic programming problem into a static one. It has been applied in guidance problems due to its ability to deal with varying velocities. The existing MPSP-based guidance methods are mainly focused on the angle constraint only. Oza [21] designed an angle-constrained guidance method for an air-to-ground missile and verified the feasibility of MPSP guidance. Maity [22] further introduced the static Lagrange multiplier in MPSP guidance, improving the computational efficiency. Refs. [23–27] improved the computational efficiency of the MPSP algorithm through further improvements and designed angle-constrained guidance methods, respectively. Refs. [21–27] verified that the MPSP algorithm is feasible for solving guidance problems online. However, during integral prediction, modeling errors will cause the accumulation of estimation errors, affecting the algorithm's performance and stability. Although the receding horizon strategy can correct some previous errors, it cannot eliminate the influence of modeling errors. Thus, the MPSP algorithm is highly dependent on the model's accuracy. To the best of the authors' knowledge, no paper has studied MPSP guidance in the presence of modeling errors so far.

Through the analysis of the ITACG method literature, we can draw the following conclusions:

(1) Compared to non-predictive guidance, predictive guidance may present better performance for unpowered missile reality applications.
(2) As one of the predictive guidance methods, MPSP-based guidance can avoid the model mismatch and thus can derive a better guidance performance, which has been verified in Refs. [21–27].
(3) The existing MPSP-based guidance methods are mainly focused on the angle constraint only. The MPSP-based ITACG methods still need further research.
(4) As an inevitable but significant factor affecting the MPSP algorithm, the modeling errors have not been considered so far.

Based on the above analysis, this paper focuses on the ITACG problems for unpowered missiles in the presence of modeling errors, trying to fill existing research gaps.

3. Problem Description

A 3-D terminal guidance scenario of a missile attacking a stationary ground target is considered in this paper, as shown in Figure 1. M is the missile whose position is (x_0, y_0, z_0). T is a stationary ground target whose position is (x_t, y_t, z_t). The missile is supposed to arrive at the target point with the desired impact path angle θ_f, impact azimuth angle ψ_{vf}, and impact time t_f.

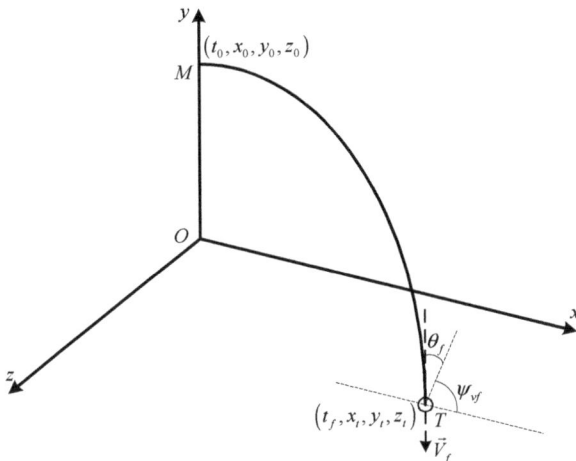

Figure 1. Terminal guidance scenario.

The dynamic model of the missile with an unknown modeling error is written as

$$\dot{X} = f(X, U, d) = \begin{bmatrix} V\cos\theta\cos\psi_v \\ V\sin\theta \\ -V\cos\theta\sin\psi_v \\ -D(\alpha,\beta) - g\sin\theta + d_x \\ \frac{L(\alpha) - g\cos\theta}{V} + d_y \\ -\frac{Z(\beta)}{V\cos\theta} + d_z \end{bmatrix}, \quad (1)$$

where $X = [x, y, z, V, \theta, \psi_v]^T$ is the state vector. x, y, and z are 3-D position coordinates of the missile. V, θ, and ψ_v are the missile's velocity, path angle, and azimuth angle, respectively. The angle of attack (AOA) α and the side slip angle (SSA) β compose the control vector $U = [\alpha, \beta]^T$. The drag acceleration D, lift acceleration L, and lateral acceleration Z are written as

$$\begin{cases} D = C_x(\alpha, \beta) q S_{ref}/m \\ L = C_y(\alpha) q S_{ref}/m \\ Z = C_z(\beta) q S_{ref}/m \end{cases}. \quad (2)$$

In Equation (2), aerodynamic coefficients C_x, C_y, and C_z are related to AOA and SSA. q is dynamic pressure, S_{ref} is reference area, m is the missile's mass. $d = [d_x, d_y, d_z]^T$ is the unknown modeling error, which may be caused by unmodeled dynamics, uncertain parameters, external disturbances, etc.

Selecting the output vector as $Y = [\begin{array}{ccccc} x & y & z & \theta & \psi_v \end{array}]^T$, the purpose of our guidance method is to determine proper control commands U, making sure $Y \to Y_d$ when $t \to t_f$, where $Y_d = [\begin{array}{ccccc} x_t & y_t & z_t & \theta_f & \psi_{vf} \end{array}]^T$.

4. CKF-MPSP Terminal Guidance Method

A CKF-MPSP guidance method is proposed and applied in the terminal guidance scenario to achieve offset-free control in the presence of modeling errors. The CKF-MPSP method comprises three parts: nominal MPSP guidance, CKF modeling error estimation, and one-step modeling-error compensation. The nominal MPSP guidance generates a baseline guidance command ignoring modeling errors. The CKF modeling error estimation generates the guessed modeling error, and the one-step modeling-error compensation introduces an error compensation term in the baseline guidance command to maintain precision and stability.

The schematic of the CKF-MPSP guidance method is shown in Figure 2. The nominal MPSP guidance method is firstly used to obtain the nominal control command U_n according to initial states X_0 and a guess value of control command U_0. Then, regarding the disturbance as an initial guess value $\hat{d}_0$, the one-step modeling-error compensation method is utilized to eliminate the modeling error's influence and generate the control command U, which is substituted into the dynamic model to update the states X. The CKF algorithm is utilized to generate the estimation of the states and the modeling error, which are denoted as $\hat{X}$ and $\hat{d}$, according to the measurement $\hat{y}$ of the global navigation satellite system (GNSS). $\hat{X}$, $\hat{d}$, and U are used as initial values to calculate the control vector in the next guidance period. Repeating the process until hit, the ITACG is achieved.

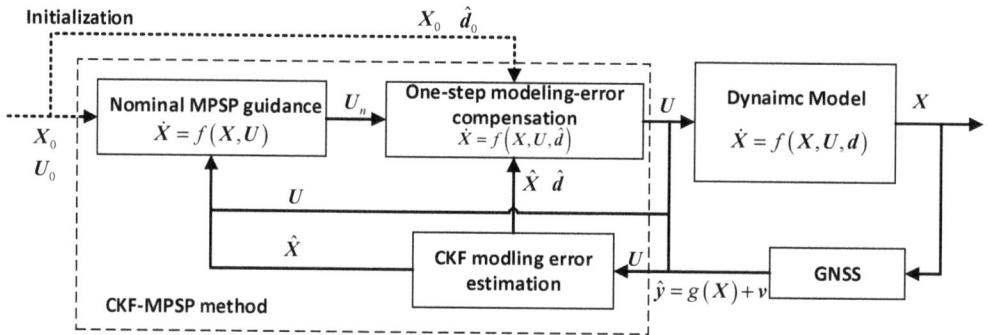

Figure 2. CKF-MPSP guidance method schematic.

4.1. Nominal MPSP Guidance

Ignoring the unknown modeling error, the MPSP method [20] is used to generate the baseline control commands. The nominal dynamic model is represented as

$$\dot{X} = f(X, U) = \begin{bmatrix} V \cos\theta \cos\psi_v \\ V \sin\theta \\ -V \cos\theta \sin\psi_v \\ -D - g\sin\theta \\ \frac{L - g\cos\theta}{V} \\ -\frac{Z}{V\cos\theta} \end{bmatrix}. \tag{3}$$

Discretizing Equation (3), the discretized dynamic model can be described as

$$\begin{cases} X_{k+1} = F_k(X_k, U_k) = X_k + h f_k(X_k, U_k) \\ Y_k = C X_k \end{cases}, \tag{4}$$

where $k = 1, 2, 3, \cdots, N$ represents the time grids and h is the simulation step. The output matrix C is shown as

$$C = \begin{bmatrix} 1 & 0 & 0 & 0 & 0 & 0 \\ 0 & 1 & 0 & 0 & 0 & 0 \\ 0 & 0 & 1 & 0 & 0 & 0 \\ 0 & 0 & 0 & 0 & 1 & 0 \\ 0 & 0 & 0 & 0 & 0 & 1 \end{bmatrix}. \tag{5}$$

After settling the desired terminal time N and the desired terminal output $Y_d = \begin{bmatrix} x_t & y_t & z_t & \theta_f & \psi_{vf} \end{bmatrix}^T$, the prediction output vector Y_N can be obtained through Runge–

Kutta integration using initial states and previous control commands. Then, the updated control commands can be obtained according to the deviation between Y_N and Y_d.

Denote the terminal output deviation as $\Delta Y_N = Y_N - Y_d$. Expanding ΔY_N at Y_d and ignoring the high-order terms, we can obtain

$$\Delta Y_N \cong dY_N = \left(\frac{\partial Y_N}{\partial X_N}\right) dX_N. \tag{6}$$

According to Equation (4), we can determine $\partial Y_N / \partial X_N = C$, and the following formula holds.

$$dX_{k+1} = \left(\frac{\partial F_k}{\partial X_k}\right) dX_k + \left(\frac{\partial F_k}{\partial U_k}\right) dU_k, \tag{7}$$

where dX_k and dU_k are the change in state and control at the k-th step, respectively.

The partial derivative of F_k with respect to X_k and U_k are shown as below:

$$\begin{aligned}\frac{\partial F_k}{\partial X_k} &= I_{6\times 6} + h \frac{\partial f_k}{\partial X_k} \\ &= I_{6\times 6} + h \begin{bmatrix} 0 & 0 & 0 & \cos\theta \cos\psi_v & -V\sin\theta\cos\psi_v & -V\cos\theta\sin\psi_v \\ 0 & 0 & 0 & \sin\theta & V\cos\theta & 0 \\ 0 & 0 & 0 & -\cos\theta\sin\psi_v & V\sin\theta\sin\psi_v & -V\cos\theta\cos\psi_v \\ 0 & 0 & 0 & 0 & -g\cos\theta & 0 \\ 0 & 0 & -\frac{L-g\cos\theta}{V^2} & & \frac{g\sin\theta}{V} & 0 \\ 0 & 0 & \frac{Z}{V^2\cos\theta} & & -\frac{Z\sin\theta}{V\cos^2\theta} & 0 \end{bmatrix}\end{aligned}, \tag{8}$$

$$\frac{\partial F_k}{\partial U_k} = h \frac{\partial f_k}{\partial U_k} = h \begin{bmatrix} 0 & 0 \\ 0 & 0 \\ 0 & 0 \\ -\frac{C_x^\alpha q S_{ref}}{m} & -\frac{C_x^\beta q S_{ref}}{m} \\ \frac{C_y^\alpha q S_{ref}}{mV} & 0 \\ 0 & -\frac{C_z^\beta q S_{ref}}{mV\cos\theta} \end{bmatrix}, \tag{9}$$

where C_x^α is the partial derivative of the drag coefficient with respect to AOA, C_y^α is the derivative of the lift coefficient with respect to AOA, C_x^β is the partial derivative of the drag coefficient with respect to SSA, and C_z^β is the derivative of the lateral coefficient with respect to SSA. They can be obtained from the aerodynamic data.

Substituting Equation (7) into Equation (6), we can obtain

$$dY_N = \left(\frac{\partial Y_N}{\partial X_N}\right) \left[\left(\frac{\partial F_{N-1}}{\partial X_{N-1}}\right) dX_{N-1} + \left(\frac{\partial F_{N-1}}{\partial U_{N-1}}\right) dU_{N-1}\right]. \tag{10}$$

In Equation (10), dX_{N-1} can be expressed as

$$dX_{N-1} = \left(\frac{\partial F_{N-2}}{\partial X_{N-2}}\right) dX_{N-2} + \left(\frac{\partial F_{N-2}}{\partial U_{N-2}}\right) dU_{N-2} \tag{11}$$

And dX_{N-2} can be further expressed by dX_{N-3} and dU_{N-3}. Repeating the above process until dX_1 and dU_1, it is clear that Equation (10) can be rewritten as

$$dY_N = AdX_1 + B_1 dU_1 + B_2 dU_2 + \cdots + B_{N-1} dU_{N-1}, \tag{12}$$

where

$$\begin{cases} A \triangleq \frac{\partial Y_N}{\partial X_N} \frac{\partial F_{N-1}}{\partial X_{N-1}} \frac{\partial F_{N-2}}{\partial X_{N-2}} \cdots \frac{\partial F_1}{\partial X_1} \\ B_k \triangleq \frac{\partial Y_N}{\partial X_N} \frac{\partial F_{N-1}}{\partial X_{N-1}} \frac{\partial F_{N-2}}{\partial X_{N-2}} \cdots \frac{\partial F_{k+1}}{\partial X_{k+1}} \frac{\partial F_k}{\partial U_k}, & k = 1, 2, \cdots, N-2 \\ B_k \triangleq \frac{\partial Y_N}{\partial X_N} \frac{\partial F_{N-1}}{\partial U_{N-1}}, & k = N-1 \end{cases} \tag{13}$$

Since the initial states are with no errors ($dX_1 = 0$), the final output error is only decided by control commands as

$$dY_N = \sum_{k=1}^{N-1} B_k dU_k. \tag{14}$$

The purpose of guidance is to find a series of control commands $U_k = U_k^0 - dU_k$ ($k = 1, 2, \cdots, N$) to make $dY_N \to 0$, where U_k^0 is the previous control history solution. It is worth noting that Equation (14) has $2 \times (N-1)$ unknowns and 5 equations. Usually, $2 \times (N-1) > 5$; thus, the solutions are not unique. To maximize guidance performance, set the following energy-optimal performance index and aim to minimize it.

$$J = \frac{1}{2} \sum_{k=1}^{N-1} \left(U_k^0 - dU_k\right)^T R_k \left(U_k^0 - dU_k\right), \tag{15}$$

where R_k is a positive definite weight coefficient matrix.

Equations (14) and (15) constitute a static optimization problem, whose solution at every time step $k = 1, 2, \cdots, N$, according to static optimization theory, is

$$U_k^* = U_k^0 - dU_k = R_k^{-1} B_k^T A_\lambda^{-1} (dY_N - b_\lambda), \tag{16}$$

where $A_\lambda \triangleq -\sum_{k=1}^{N-1} B_k R_k^{-1} B_k^T, b_\lambda \triangleq \sum_{k=1}^{N-1} B_k U_k^0$.

4.2. Modeling Error Estimation Based on CKF

The MPSP guidance method highly relies on modeling accuracy because of the integral prediction. However, the realistic model inevitably has unknown modeling errors or external disturbances. It has been pointed out in the literature [28] that, in the presence of model mismatch, the MPSP method cannot realize the desired terminal states. Estimating and compensating for modeling errors are common ways to achieve offset-free terminal state control. This section uses the CKF algorithm to estimate states and modeling errors simultaneously for subsequent compensation.

To estimate the modeling errors, consider them as constants and extend them to states. The dynamic model (1) can be rewritten as

$$\dot{X}^E = f_E\left(X^E\right) = \begin{bmatrix} V \cos\theta \cos\psi_v \\ V \sin\theta \\ -V \cos\theta \sin\psi_v \\ -D - g\sin\theta + d_x \\ \frac{L - g\cos\theta}{V} + d_y \\ -\frac{Z}{V\cos\theta} + d_z \\ 0 \\ 0 \\ 0 \end{bmatrix} + w, \tag{17}$$

where $X^E = [x, y, z, V, \theta, \psi_v, d_x, d_y, d_z]^T$ is the expansion state vector, w is Gaussian-distributed process noise, and $E[ww^T] = Q$.

During flight, GNSS measures the missile's motion in real time. So, the measurement equations can be denoted as

$$\hat{y}^E = X^E + v, \tag{18}$$

where v is Gaussian-distributed measurement noise and $E[vv^T] = R$. $\hat{y}^E = [\hat{x}, \hat{y}, \hat{z}, \hat{V}, \hat{\theta}, \hat{\psi}_v, \hat{d}_x, \hat{d}_y, \hat{d}_z]^T$ is the expansion output vector.

Discretizing Equations (17) and (18), we can obtain the nonlinear filter model:

$$\begin{cases} X_k^E = F_E(X_{k-1}^E) + w_{k-1} \\ Y_k^E = X_k^E + v_{k-1} \end{cases}, \quad (19)$$

where $F_E(X_{k-1}^E) = X_{k-1}^E + hf_E(X_{k-1}^E)$.

The CKF algorithm consists of two procedures: Time Update and Measurement Update [29]. Combined with the filter model, the CKF algorithm process is shown below.

4.2.1. Time Update

Assume at time k that the posterior probability density function $p(X_{k-1}^E|y_{k-1}) = N(\hat{X}_{k-1}^E, P_{k-1})$ is known. Denote the Cholesky factorization of the error covariance P_{k-1} as S_{k-1}.

$$P_{k-1} = S_{k-1}S_{k-1}^T. \quad (20)$$

Calculate the cubature points $\chi_{k-1}^{(i)}$ based on the third-degree cubature rule:

$$\chi_{k-1}^{(i)} = \hat{X}_{k-1}^E + S_{k-1}\xi_i, \quad i = 1, 2, \cdots, 2n, \quad (21)$$

where n is the dimension of states. $\xi_i = \sqrt{n}[1]_i$ is the basic cubature point set. The point set $[1]$ is defined as

$$[1] = \underbrace{\left\{ \begin{bmatrix} 1 \\ 0 \\ \vdots \\ 0 \end{bmatrix}_{n \times 1}, \begin{bmatrix} 0 \\ 1 \\ \vdots \\ 0 \end{bmatrix}, \cdots, \begin{bmatrix} 0 \\ 0 \\ \vdots \\ 1 \end{bmatrix}, \begin{bmatrix} -1 \\ 0 \\ \vdots \\ 0 \end{bmatrix}, \begin{bmatrix} 0 \\ -1 \\ \vdots \\ 0 \end{bmatrix}, \cdots, \begin{bmatrix} 0 \\ 0 \\ \vdots \\ -1 \end{bmatrix} \right\}}_{2n}, \quad (22)$$

and $[1]_i$ represents the i-th column vector in $[1]$.

Calculate the one-step prediction at time k and its error covariance:

$$\chi_{k|k-1}^{*(i)} = F_E\left(\chi_{k-1}^{(i)}\right), \quad (23)$$

$$\hat{X}_{k|k-1}^E = \frac{1}{2n}\sum_{i=1}^{2n} \chi_{k|k-1}^{*(i)}, \quad (24)$$

$$P_{k|k-1} = \frac{1}{2n}\sum_{i=1}^{2n} \left[\chi_{k|k-1}^{*(i)} - \hat{X}_{k|k-1}^E\right]\left[\chi_{k|k-1}^{*(i)} - \hat{X}_{k|k-1}^E\right]^T + Q_{k-1}. \quad (25)$$

4.2.2. Measurement Update

Calculate the cubature points for Measurement Update, and then calculate measurement prediction:

$$P_{k|k-1} = S_{k|k-1}S_{k|k-1}^T, \quad (26)$$

$$\chi_{k|k-1}^{(i)} = \hat{X}_{k|k-1}^E + S_{k|k-1}\xi_i, \quad i = 1, 2, \cdots, 2n, \quad (27)$$

$$\hat{y}_{k|k-1}^E = \frac{1}{2n}\sum_{i=1}^{2n} \chi_{k|k-1}^{(i)}. \quad (28)$$

Calculate the innovation covariance matrix P_{xy} and the cross-covariance matrix P_{yy}:

$$P_{xy} = \frac{1}{2n}\sum_{i=1}^{2n}\left[\chi_{k|k-1}^{(i)} - \hat{X}_{k|k-1}^{E}\right]\left[\chi_{k|k-1}^{(i)} - \hat{y}_{k|k-1}^{E}\right]^{T}, \quad (29)$$

$$P_{yy} = \frac{1}{2n}\sum_{i=1}^{2n}\left[\chi_{k|k-1}^{(i)} - \hat{y}_{k|k-1}^{E}\right]\left[\chi_{k|k-1}^{(i)} - \hat{y}_{k|k-1}^{E}\right]^{T} + R_{k}. \quad (30)$$

Calculate the Kalman gain:

$$K_k = P_{xy}P_{yy}^{-1}. \quad (31)$$

Estimate the updated state and the corresponding error covariance:

$$\hat{X}_{k}^{E} = \hat{X}_{k|k-1}^{E} + K_{k}\left(y_{k}^{E} - \hat{y}_{k|k-1}^{E}\right), \quad (32)$$

$$P_k = P_{k|k-1} - K_k P_{yy} K_k^T. \quad (33)$$

4.3. One-Step Modeling-Error Compensation

After estimating the modeling errors, compensate for the effect of modeling errors on the system by attaching an additional control term ΔU_k to the MPSP optimal command U_k^* at time k. Denoting the estimation modeling error vector as $\hat{d}_k = \left[\hat{d}_{xk}, \hat{d}_{yk}, \hat{d}_{zk}\right]^T$, the disturbed system model can be described as

$$\begin{cases} \tilde{X}_{k+1} = \tilde{X}_k + h f_k\left(\tilde{X}_k, \tilde{U}_k, \hat{d}_k\right) \\ \tilde{Y}_{k+1} = C\tilde{X}_{k+1} \end{cases}. \quad (34)$$

At time k, denote Y_{k+1} as the output vector obtained by a one-step calculation with current state X_k and the MPSP control command U_k^*. The objective of modeling-error compensation is to generate a modified control command $\tilde{U}_k = U_k^* + \Delta U_k$ making $\tilde{Y}_{k+1} \to Y_{k+1}$.

Denote the output error as

$$\Delta Y_{k+1} = \tilde{Y}_{k+1} - Y_{k+1}. \quad (35)$$

Substituting Equations (4) and (34) into (35) and because of $\tilde{X}_k = X_k$, we can obtain

$$\Delta Y_{k+1} = hC\left[f_k(X_k, U_k^* + \Delta U_k, \hat{d}_k) - f_k(X_k, U_k^*)\right]. \quad (36)$$

Expanding $f_k\left(X_k, U_k^* + \Delta U_k, \hat{d}_k\right)$ and ignoring high-order terms, we can obtain

$$f_k(X_k, U_k^* + \Delta U_k, \hat{d}_k) \cong f_k(X_k, U_k^*) + \frac{\partial f_k}{\partial U_k}\Delta U_k + \frac{\partial f_k}{\partial d_k}\hat{d}_k. \quad (37)$$

Substituting Equation (37) into (36), we can obtain

$$\Delta Y_{k+1} = hC\left[\frac{\partial f_k}{\partial U_k}\Delta U_k + \frac{\partial f_k}{\partial d_k}\hat{d}_k\right]. \quad (38)$$

The compensation term is desired to make the output error zero. According to Equation (38), we can obtain the desired additional control term as

$$\Delta U_k = -\left(C\frac{\partial f_k}{\partial U_k}\right)^{-1}\left(C\frac{\partial f_k}{\partial d_k}\hat{d}_k\right). \quad (39)$$

The modified control command at time k is

$$\tilde{U}_k = U_k^* + \Delta U_k. \qquad (40)$$

4.4. CKF-MPSP Terminal Guidance Process

Considering the above, the CKF-MPSP guidance process is summarized below (Algorithm 1).

Algorithm 1: CKF-MPSP Terminal Guidance Scheme.

INPUT: current time k, desired terminal time N, desired terminal output Y_d, current states X_k, guessed control command $\left[U_k^0, U_{k+1}^0, \cdots, U_{N-1}^0\right]$, estimated modeling error $\hat{d}_k$, CKF initial values $\hat{X}_{k-1}^E$, $\hat{P}_{k-1}$.

1: **while** current time k is no larger than N, **do**
2:    **while** terminal output deviation ΔY_N is larger than tolerance value ε, **do**
3:      predict the terminal output vector Y_N through Runge-Kutta integration with the nominal dynamic model (4), the current states X_k and the guessed control command $\left[U_k^0, U_{k+1}^0, \cdots, U_{N-1}^0\right]$.
4:      calculate terminal output deviation $\Delta Y_N = Y_N - Y_d$.
5:      calculate matrices $\left[B_k, B_{k+1}, \cdots, B_{N-1}\right]$ according to Equation (13).
6:      calculate the optimal control command $\left[U_k^*, U_{k+1}^*, \cdots, U_{N-1}^*\right]$ according to Equation (16).
7:      take $\left[U_{k+1}^*, \cdots, U_{N-1}^*\right]$ as the new guessed control command.
8:    **end while**.
9:    calculate the one-step output $\tilde{Y}_{k+1}$ with the disturbed system model (34) in presence of the estimated modeling error $\hat{d}_k$
10:    calculate the modified control command $\tilde{U}_k$ at time k, according to Equations (39) and (40).
11:    substitute $\tilde{U}_k$ into the realistic dynamic model (1) and obtain the updated state X_{k+1}.
12:    estimate the filter state $\hat{X}_k^E$, error covariance $\hat{P}_k$, and modeling error $\hat{d}_{k+1}$, using CKF algorithm (20)~(33).
13:    time update, $k = k + 1$.
14: **end while**

Remark 1. *The MPSP guidance method takes the desired impact time as the terminal time N of static planning. The target's position and impact angles are regarded as desired terminal output Y_d. Making the terminal output deviation ΔY_N no larger than tolerance value ε by iterating U^*, the impact time and angle constraints can be satisfied simultaneously.*

Remark 2. *The MPSP guidance method relies on initial guessed control commands $\left[U_k^0, U_{k+1}^0, \cdots, U_{N-1}^0\right]$. The guessed control commands are quickly generated through some simple guidance laws in common. In this paper, the traditional PN guidance law is used to obtain the initial guessed control commands. Usually, the impact time of PN, represented by the symbol N_P, is different from the desired time N. For the MPSP algorithm, N_P must be no less than N, so the outputs at time N can be predicted. In this paper, a protection mechanism is introduced to make sure the MPSP algorithm normally runs even if N_P is less than N: Let $U_k^0 = U_{N_P-1}^0$, for $N_P - 1 < k \leq N - 1$.*

5. Simulations and Results

In this section, several numerical simulations are carried out to evaluate the performance of the proposed terminal guidance method in the presence of modeling errors. A three-dimensional guidance scenario of a missile attacking a stationary target is constructed. The initial simulation conditions are listed below (Table 1):

Table 1. Simulation initial conditions.

Parameters	Values
Missile's initial velocity V	200 m/s
Missile's initial path angle θ	0°
Missile's initial azimuth angle ψ_v	0°
Missile's initial position (x, y, z)	(0 m, 4000 m, 0 m)
Target's position (x_t, y_t, z_t)	(5000 m, 0 m, 1000 m)

Besides achieving precision arrival, the impact time and impact angles are also required. The desired impact time is settled as 35 s, and the desired terminal path angle and azimuth angle are settled as $-80°$ and $-50°$, respectively. The dynamic modeling errors in Equation (1) are settled as $d_x = -0.4\sin(t/400)\cos(t/400)$, $d_y = 1$, and $d_z = 2.5\cos(\pi t/200)$. The measurement errors from the GNSS system are assumed to be normally distributed. The position error, velocity error, and acceleration error are settled to be 10 m (3σ), 1 m/s (3σ), and 0.1 m/s² (3σ), respectively.

For the MPSP algorithm, the guessed control commands $[u_k^0, u_{k+1}^0, \cdots, u_{N-1}^0]$ are needed. The traditional PN [30] is used to produce the initial values of $[u_k^0, u_{k+1}^0, \cdots, u_{N-1}^0]$ in this paper, and the navigation ratio is settled as 6. In addition, a comparison with the MPSP guidance method presented in [21] is provided to validate the superiority of the method. The end condition for PN simulation is that the missile reaches the target. And the end conditions for the other two simulations are that the simulation times reach the desired impact time, which is 35 s on this occasion. For MPSP and CKF-MPSP methods, the tolerance value vector is settled as $\varepsilon = [1\text{m}, 1\text{m}, 1\text{m}, 0.1°, 0.1°]^T$. The simulation results are shown below.

To evaluate the guidance accuracy, some crucial parameters of the three methods shown in Figure 3 are provided in Tables 2 and 3, which include the terminal miss distance, terminal velocity, terminal path angle, terminal azimuth angle, and impact time.

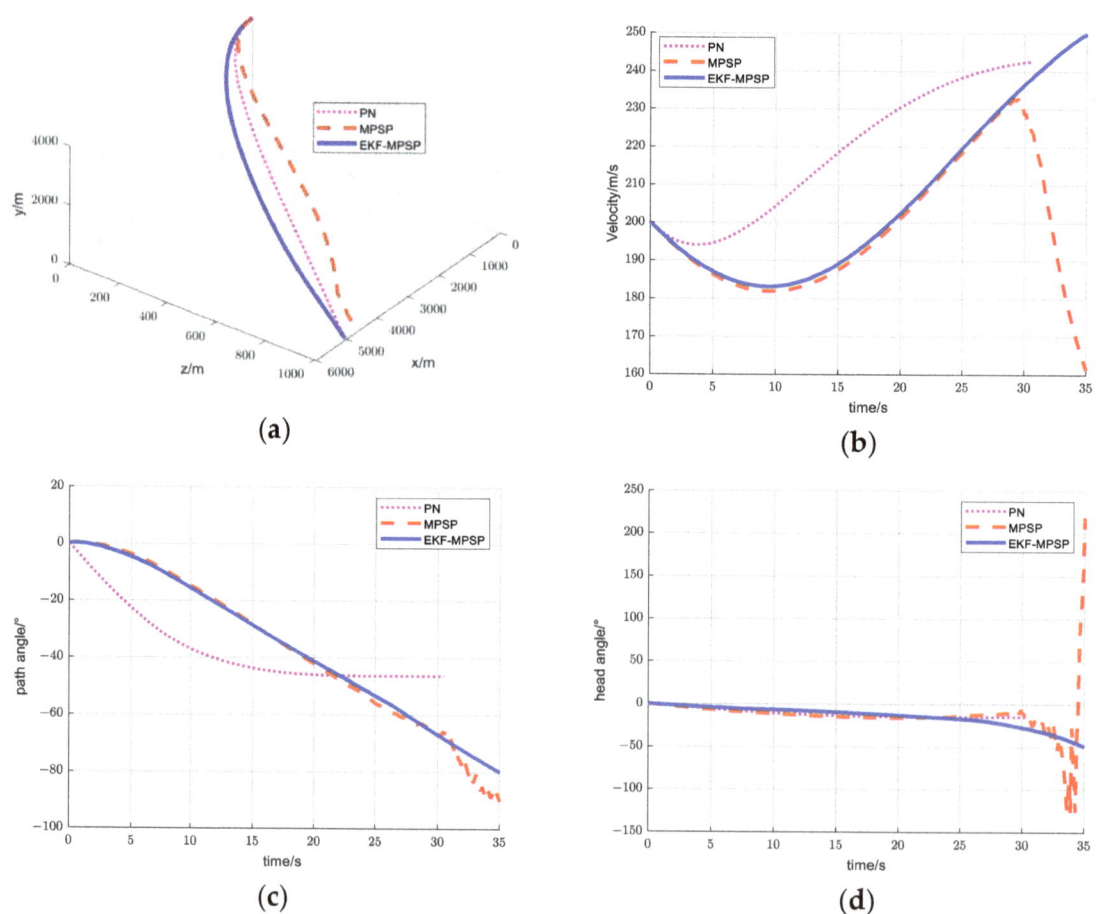

Figure 3. Cont.

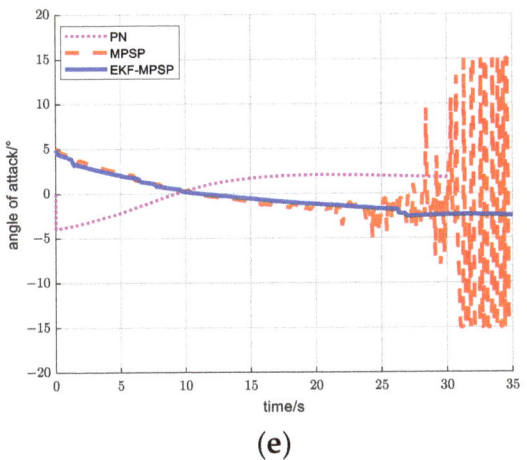
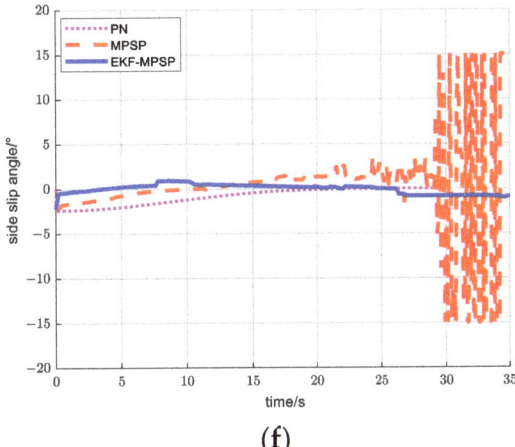

Figure 3. Comparative simulation results of three methods: (**a**) three-dimensional trajectory; (**b**) missile velocity profiles; (**c**) path angle profiles; (**d**) azimuth angle profiles; (**e**) angle-of-attack profiles; and (**f**) side slip angle profiles.

Table 2. Simulation results of PN method.

Method	Miss Distance (m)	Velocity (m/s)	Path Angle (°)	Azimuth Angle (°)	Impact Time (s)
PN	0.26	242.56	−46.42	−15.23	30.5

Table 3. Simulation results of MPSP and CKF-MPSP methods at the desired impact time (35 s).

Method	Miss Distance (m)	Velocity (m/s)	Path Angle (°)	Azimuth Angle (°)
MPSP	289.73	161.18	−90.24	218.31
CKF-MPSP	0.29	249.90	−79.99	−50.00

It is obvious that the CKF-MPSP method has good accuracy for miss distance while strictly constraining the terminal path angle, azimuth angle, and impact time. The PN method leads to the minimum miss distance. However, it cannot consider impact time and angle constraints. Because of the dynamic modeling errors, the MPSP method cannot find a feasible solution, which leads to a significant guidance error. Affected by modeling errors, the MPSP method's velocity is lower than the CKF-MPSP's at every identical moment, which is also the main reason for the MPSP method not reaching the destination. The CKF-MPSP method estimates the modeling errors with high accuracy. Referring to Figure 4, the estimation error of modeling errors is no larger than 0.04 m/s^2. The influence of modeling errors can be reduced by compensating for the nominal command acceleration. The simulation results illustrate the effectiveness and superiority of the CKF-MPSP method in the presence of modeling errors.

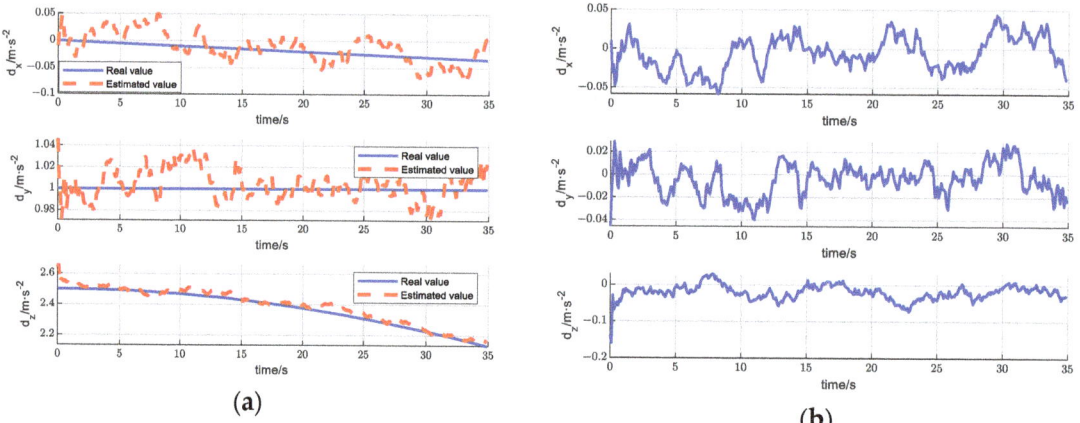

Figure 4. CKF estimation results: (**a**) CKF estimation of modeling errors and (**b**) estimation error of modeling errors.

Based on the above simulation, the influence of atmospheric density deviation, aerodynamic parameter deviation, and the random variation in the dynamic modeling errors are further considered. The deviations are assumed to be normally distributed, and their values are 10% (3σ). The results of 200 Monte Carlo simulations are as follows.

The key indexes in Figure 5 is summarized in Table 4. According to Table 4, the miss distances are no larger than 5.56 m and the average impact angle errors are $0.055°$ and $0.077°$, respectively. The missile maintains high guidance accuracy in the presence of disturbances. The simulation results illustrate that the proposed guidance method has strong robustness.

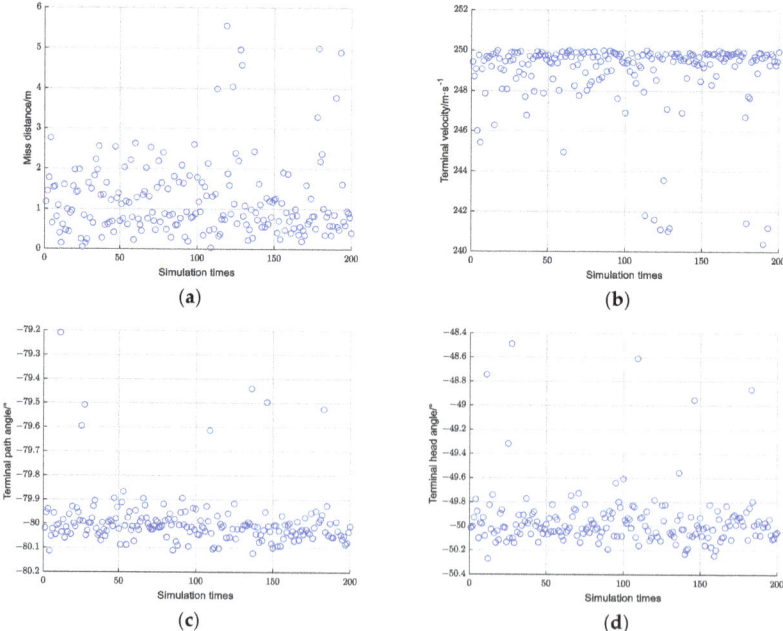

Figure 5. Monte Carlo simulation results: (**a**) terminal miss distance; (**b**) terminal velocity; (**c**) path angle profiles; and (**d**) azimuth angle profiles.

Table 4. Monte Carlo simulation results.

Method	Miss Distance (m)	Velocity (m/s)	Path Angle (°)	Azimuth Angle (°)
Average	1.21	248.92	−79.95	−49.93
Maximum value	5.56	250.11	−79.21	−48.49
Minimum value	0.03	240.40	−80.12	−50.26
Standard error	0.95	1.84	0.16	0.34

6. Conclusions

ITACG is a vital field for missiles because it can improve destructive effects and hit weak parts of time-sensitive targets, and make it possible for multiple missiles to attack a target simultaneously. Studying ITACG can effectively enhance the combat effectiveness of missiles.

In this paper, an ITACG is proposed based on the MPSP algorithm. By taking the desired impact time and angles as terminal conditions, the guidance method can satisfy these constraints simultaneously. Furthermore, to eliminate the influence of modeling errors on prediction, the CKF algorithm is used for error estimation, and a compensation scheme is designed. The proposed guidance method considers the missile's dynamic model instead of a constant-velocity model. Meanwhile, the modeling errors are estimated and compensated. Thus, this method is more practically significant. A terminal guidance scenario is settled, and the PN method, MPSP method, and CKF-MPSP method are used for simulation in the presence of modeling errors. According to the simulation results, the CKF-MPSP method can achieve impact time and angle constraint guidance, and maintain high accuracy within the influence of modeling errors. Furthermore, the Monte Carlo simulation is conducted, considering the influence of atmospheric density deviations, aerodynamic parameter deviations, and random variations in the dynamic modeling errors. According to the simulation results, the miss distances are no larger than 5.56 m, and the average impact angle errors are 0.055° and 0.077°, respectively. The missile maintains high guidance accuracy in the presence of disturbances. Comprehensively, the simulation results illustrate that the proposed CKF-MPSP guidance method has high precision and strong robustness.

It should be acknowledged that this article still has some limitations. The research of this paper is mainly focused on the ITACG against stationary targets. For moving targets, it is also necessary to introduce their motion models into the guidance method. However, the accurate estimation of the moving targets' motion is still a difficult problem, because of insufficient target information and potential maneuvering. Motion model estimation and MPSP-based guidance for moving targets remain to be researched in the future.

Author Contributions: Conceptualization, Z.X. and J.P.; methodology, Z.X.; software, Z.X.; validation, J.P. and C.W.; writing—original draft, Z.X.; writing—review and editing, J.P., C.W. and Y.G. All authors have read and agreed to the published version of the manuscript.

Funding: This research was funded by the National Natural Science Foundation of China, grant number U2241215.

Data Availability Statement: Not applicable.

Conflicts of Interest: The authors declare no conflict of interest.

References

1. Yu, H.; Dai, K.; Li, H.; Zou, Y.; Ma, X.; Ma, S.; Zhang, H. Three-dimensional adaptive fixed-time cooperative guidance law with impact time and angle constraints. *Aerosp. Sci. Technol.* **2022**, *123*, 107450. [CrossRef]
2. Zhang, S.; Guo, Y.; Liu, Z.; Wang, S.; Hu, X. Finite-time cooperative guidance strategy for impact angle and time control. *IEEE Trans. Aerosp. Electron. Syst.* **2020**, *57*, 806–819. [CrossRef]
3. Chen, Z.; Chen, W.; Liu, X.; Cheng, J. Three-dimensional fixed-time robust cooperative guidance law for simultaneous attack with impact angle constraint. *Aerosp. Sci. Technol.* **2021**, *110*, 106523. [CrossRef]
4. Ma, F.; Wu, Y.; Wang, S.; Yang, X.; Hua, Y. Three-dimensional adaptive fixed-time guidance law against maneuvering targets with impact angle constraints and control input saturation. *Trans. Inst. Meas. Control* **2022**, *44*, 1579–1598. [CrossRef]

5. Lin, M.; Ding, X.; Wang, C.; Liang, L.; Wang, J. Three-dimensional fixed-time cooperative guidance law with impact angle constraint and prespecified impact time. *IEEE Access* **2021**, *9*, 29755–29763. [CrossRef]
6. Wang, Z.; Fang, Y.; Fu, W.; Wu, Z.; Wang, M. Cooperative guidance laws against highly maneuvering target with impact time and angle. *Proc. Inst. Mech. Eng. Part G J. Aerosp. Eng.* **2022**, *236*, 1006–1016. [CrossRef]
7. Jing, L.; Wei, C.; Zhang, L.; Cui, N. Cooperative guidance law with predefined-time convergence for multimissile systems. *Math. Probl. Eng.* **2021**, *2021*, 9940240. [CrossRef]
8. Chen, X.; Wang, J. Optimal control based guidance law to control both impact time and impact angle. *Aerosp. Sci. Technol.* **2019**, *84*, 454–463. [CrossRef]
9. Kim, T.H.; Lee, C.H.; Jeon, I.S.; Tahk, M.J. Augmented polynomial guidance with impact time and angle constraints. *IEEE Trans. Aerosp. Electron. Syst.* **2013**, *49*, 2806–2817. [CrossRef]
10. Zhao, Y.; Sheng, Y.; Liu, X. Trajectory reshaping based guidance with impact time and angle constraints. *Chin. J. Aeronaut.* **2016**, *29*, 984–994. [CrossRef]
11. Kang, S.; Tekin, R.; Holzapfel, F. Generalized impact time and angle control via look-angle shaping. *J. Guid. Control Dyn.* **2019**, *42*, 695–702. [CrossRef]
12. Hou, Z.; Liu, L.; Wang, Y. Time-to-go estimation for terminal sliding mode based impact angle constrained guidance. *Aerosp. Sci. Technol.* **2017**, *71*, 685–694. [CrossRef]
13. Hou, Z.; Yang, Y.; Liu, L.; Wang, Y. Terminal sliding mode control based impact time and angle constrained guidance. *Aerosp. Sci. Technol.* **2019**, *93*, 105142. [CrossRef]
14. Chen, X.; Wang, J. Two-stage guidance law with impact time and angle constraints. *Nonlinear Dyn.* **2019**, *95*, 2575–2590. [CrossRef]
15. Zhang, Y.; Ma, G.; Liu, A. Guidance law with impact time and impact angle constraints. *Chin. J. Aeronaut.* **2013**, *26*, 960–966. [CrossRef]
16. Yan, X.; Zhu, J.; Kuang, M.; Yuan, X. A computational-geometry-based 3-dimensional guidance law to control impact time and angle. *Aerosp. Sci. Technol.* **2020**, *98*, 105672. [CrossRef]
17. Majumder, K.; Kumar, S.R. Sliding mode–based simultaneous control of impact angle and impact time. *Proc. Inst. Mech. Eng. Part G J. Aerosp. Eng.* **2022**, *236*, 1269–1281. [CrossRef]
18. Liu, X.; Li, G. Adaptive Sliding Mode Guidance with Impact Time and Angle Constraints. *IEEE Access* **2020**, *8*, 26926–26932. [CrossRef]
19. Wang, Z.; Hu, Q.; Han, T.; Xin, M. Two-Stage Guidance Law with Constrained Impact via Circle Involute. *IEEE Trans. Aerosp. Electron. Syst.* **2021**, *57*, 1301–1316. [CrossRef]
20. Padhi, R.; Kothari, M. Model predictive static programming: A computationally efficient technique for suboptimal control design. *Int. J. Innov. Comput. Inf. Control* **2009**, *5*, 399–411.
21. Oza, H.B.; Padhi, R. Impact-Angle-Constrained Suboptimal Model Predictive Static Programming Guidance of Air-to-Ground Missiles. *J. Guid. Control Dyn.* **2012**, *35*, 153–164. [CrossRef]
22. Maity, A.; Oza, H.B.; Padhi, R. Generalized model predictive static programming and angle-constrained guidance of air-to-ground missiles. *J. Guid. Control Dyn.* **2014**, *37*, 1897–1913. [CrossRef]
23. Mondal, S.; Padhi, R. Angle-constrained terminal guidance using quasi-spectral model predictive static programming. *J. Guid. Control Dyn.* **2018**, *41*, 783–791. [CrossRef]
24. He, X.; Chen, W.; Yang, L. Suboptimal impact-angle-constrained guidance law using linear pseudospectral model predictive spread control. *IEEE Access* **2020**, *8*, 102040–102050. [CrossRef]
25. Zhou, C.; Yan, X.; Tang, S. Generalized quasi-spectral model predictive static programming method using Gaussian quadrature collocation. *Aerosp. Sci. Technol.* **2020**, *106*, 106134. [CrossRef]
26. Mondal, S.; Padhi, R. Constrained Quasi-Spectral MPSP With Application to High-Precision Missile Guidance with Path Constraints. *J. Dyn. Syst. Meas. Control* **2021**, *143*, 031001. [CrossRef]
27. Liu, X.; Li, S.; Xin, M. Pseudospectral Convex Optimization Based Model Predictive Static Programming for Constrained Guidance. *IEEE Trans. Aerosp. Electron. Syst.* **2023**, *59*, 2232–2244. [CrossRef]
28. Mathavaraj, S.; Padhi, R. Unscented MPSP for optimal control of a class of uncertain nonlinear dynamic systems. *J. Dyn. Syst. Meas. Control* **2019**, *141*, 65001. [CrossRef]
29. Arasaratnam, I.; Haykin, S. Cubature Kalman filters. *IEEE Trans. Autom. Control* **2009**, *54*, 1254–1269. [CrossRef]
30. Becker, K. Closed-form solution of pure proportional navigation. *IEEE Trans. Aerosp. Electron. Syst.* **1990**, *26*, 526–533. [CrossRef]

Disclaimer/Publisher's Note: The statements, opinions and data contained in all publications are solely those of the individual author(s) and contributor(s) and not of MDPI and/or the editor(s). MDPI and/or the editor(s) disclaim responsibility for any injury to people or property resulting from any ideas, methods, instructions or products referred to in the content.

Article

Hybrid Attitude Saturation and Fault-Tolerant Control for Rigid Spacecraft without Unwinding

Jun Ma, Zeng Wang * and Chang Wang

Xi'an Key Laboratory of Intelligence, Xi'an Technological University, Xi'an 710021, China; majun@xatu.edu.cn (J.M.); wangchang86@st.xatu.edu.cn (C.W.)
* Correspondence: wangzeng@xatu.edu.cn

Abstract: This paper tackles the saturation and fault-tolerant attitude tracking problem without unwinding for rigid spacecraft with external disturbances and partial loss of actuator effectiveness faults. A hybrid saturation and fault-tolerant attitude control (HSFC) is proposed. The Lyapunov method is employed to prove that the tracking errors of the spacecraft system tend to the equilibrium point asymptotically with HSFC. The advantages of the HSFC are that it is fault-tolerant, anti-unwinding and explicitly upper bounded a priori which means that both actuator saturation and the unwinding phenomenon can be avoided. Simulations verify the effectiveness of the proposed approach.

Keywords: spacecraft attitude control; saturation control; unwinding; fault tolerant

MSC: 93D20

1. Introduction

In the last few decades, the attitude control of rigid spacecraft has attracted extensive attention and several elegant attitude control strategies for rigid spacecraft have been proposed. More specifically, the authors of [1] using the passivity theory develop an adaptive control scheme for the attitude control of rigid spacecraft. In [2], an adaptive finite time nonsingular terminal sliding mode attitude tracking control (AFNTSMC) scheme is presented for uncertain rigid spacecraft. In [3], the authors propose a simple non-singular terminal sliding mode control (NTSMC) to obtain high precision and robust finite-time bounded attitude tracking for rigid spacecraft with finite-time stability. In [4], a new integral sliding mode control integrating the bi-limit homogeneous theory is explored to obtain fixed-time stability for rigid spacecraft attitude tracking. Recently, the authors of [5] exploit the predefined-time guaranteed performance takeover control for non-cooperative spacecraft.

But, these aforementioned controls are formulated with the assumption that the actuators could supply any requested torque for the attitude control of spacecraft. In a practical scenario, when the requested control torque exceeds the maximum value that the actuator can supply, the performance of the spacecraft system cannot be guaranteed and even leads to instability. Obviously, it is more unrealistic to design a robust control strategy under the above assumption [6,7]. Recognizing this drawback, several approximate solutions that take into account actuator constraints have been proposed. Particularly, the authors of [8] propose a continuous globally robust attitude saturation control for spacecraft in the presence of parametric uncertainty and external disturbances. In [9], a nonlinear backstepping attitude saturation control integrating the inverse tangent-based tracking function and a family of augmented Lyapunov functions is exploited to achieve attitude maneuver of rigid spacecraft. In [10], an adaptive saturation attitude tracking control is designed for rigid spacecraft with unknown system parameters and disturbance. In [11], two very simple saturated PD (SPD) controllers are developed for rigid spacecraft to obtain global asymptotic stabilization. Subsequently, velocity-free asymptotic attitude stabilization control is introduced for rigid spacecraft in the presence of actuator constraints [12]. In [13],

Citation: Ma, J.; Wang, Z.; Wang, C. Hybrid Attitude Saturation and Fault-Tolerant Control for Rigid Spacecraft without Unwinding. *Mathematics* **2023**, *11*, 3431. https://doi.org/10.3390/math11153431

Academic Editors: Haizhao Liang, Jianying Wang and Chuang Liu

Received: 17 July 2023
Revised: 2 August 2023
Accepted: 3 August 2023
Published: 7 August 2023

Copyright: © 2023 by the authors. Licensee MDPI, Basel, Switzerland. This article is an open access article distributed under the terms and conditions of the Creative Commons Attribution (CC BY) license (https://creativecommons.org/licenses/by/4.0/).

a unified formulation of simple but effective SPD control is proposed for asymptotic stabilization of spacecraft in the presence of actuator constraints. In [14], a simple single saturated PD (SSPD) control is proposed for spacecraft stabilization. In [15], a saturated output feedback finite-time proportional–derivative control is developed for spacecraft subject to actuator constraints and attitude measurements only.

In spite of the above-mentioned schemes addressing the attitude saturation problem of the spacecraft, they do not consider the actuator faults. It should be pointed out that actuator faults of spacecraft may dramatically degrade the attitude tracking performance [16,17]. To eliminate this weakness, a fault-tolerant technique is added to the spacecraft attitude control scheme to improve the safety and accuracy of the attitude tracking. Recognizing this benefit, several effective fault-tolerant control schemes for spacecraft attitude control have been developed to compensate actuator failure. The authors of [18] develop an adaptive robust fault-tolerant control to tackle the spacecraft attitude tracking problem. In [19], an adaptive fault-tolerant control with fast transient is proposed to address spacecraft attitude tracking. The authors of [20] introduce a fault-tolerant on-line control to solve the spacecraft attitude tracking with actuator failure. In [21], a fixed-time fault-tolerant attitude tracking control is explored for rigid spacecraft described by the unit quaternion subject to model uncertainties, external disturbances and actuator faults. In [22], based on the fixed-time disturbance observer, a quantized fixed-time control is introduced to obtain attitude stabilization. In [23], an incremental nonlinear control technology is used to simplify the attitude control system with a synthetic uncertainty or fault term.

For the unit-quaternion representation, although it is a global nonsingularity, it has the weakness of the unwinding phenomenon [24]. In comparison with the almost 'global' stability in the above quaternion-based controls, the hysteresis-based hybrid attitude control can ensure that the global stability of the spacecraft system is obtained. Recognizing these advantages, several hysteresis-based hybrid attitude control schemes have been exploited. The authors of [25] propose a quaternion-based hybrid feedback scheme to address global attitude stabilization without the angular velocity measurement. The authors of [26] present a smooth control system, which can provide almost semi-global exponential stability. The authors of [27] introduce a hybrid certainty equivalence controller scheme with a hybrid observer for the rigid spacecraft with only quaternion measurement. More recently, in [28], a global finite-time attitude control based on the hybrid control technique is designed to solve the attitude tracking of a rigid body using a quaternion description. In [29], a saturated hybrid output feedback PD plus (SHOPD+) scheme with attitude measurements only is developed to achieve global stability for rigid spacecraft subject to the actuator limit. Furthermore, a velocity-free saturated hybrid proportional–derivative (PD) plus (PD+) control is constructed to achieve global finite-time attitude tracking for spacecraft [30]. The authors of [31] present a novel anti-unwinding finite-time attitude tracking control law with a designed control signal which works within a known actuator-magnitude constraint using a continuous nonsingular fast terminal sliding mode (NFTSM) concept.

In this paper, a simple hybrid attitude saturation and fault-tolerant control is proposed to address the spacecraft attitude tracking problem subject to external disturbances and partial loss of actuator effectiveness faults. An adaptive hybrid robust saturation control is developed to obtain global stability which means the tracking errors tend to the equilibrium point asymptotically without the unwinding phenomenon. In comparison with the existing saturation attitude controls of spacecraft in [29,30], the proposed control can tackle actuator faults. Compared with the available fault-tolerant control schemes for spacecraft in [31], the proposed control can remove the possibility of degraded or unpredictable motion and actuator failure due to excessive torque input levels by selecting control gains a priori. Advantages of the proposed control include anti-unwinding, global stability, control constraint, fault-tolerance and robustness. Simulations are performed on the spacecraft to verify the effectiveness performance of the developed HSFC.

Throughout this paper, notations $\lambda_m(K)$ and $\lambda_M(K)$ are utilized to denote the smallest and largest eigenvalues, respectively, of a symmetric positive-definite bounded matrix K.

We use $\|x\| = \sqrt{x^T x}$ to define the norm of a vector $x \in R^n$ and the corresponding induced norm $\|K\| = \sqrt{\lambda_M(K^T K)}$ is used to define the norm of a matrix K, and I_3 denotes an $R^{3\times 3}$ identity matrix.

The framework of this paper is organized as follows. The preliminaries are given in Section 2. The control design including hybrid system and controller formulation is presented in Section 3. In Section 4, asymptotic stability analysis is given. In Section 5, numerical simulations are illustrated to verify the effectiveness performance of the proposed approach. Finally, the conclusion is presented in Section 6.

2. Preliminaries

2.1. Spacecraft Model and Properties

The attitude kinematics and dynamics of a rigid spacecraft are formulated as [2,32]:

$$\begin{cases} \dot{q}_v = \frac{1}{2}(q_4 I_3 + q_v^\times)\omega, \\ \dot{q}_4 = -\frac{1}{2}q_v^T \omega. \end{cases} \quad (1)$$

$$J\dot{\omega} = -\omega^\times J\omega + u + d. \quad (2)$$

where a unit quaternion $q \in \bar{S}^3 = \{x \in R^4 : x^T x = 1\}$ is used to describe the attitude orientation of the spacecraft in the body frame with respect to an inertial frame, and $\bar{S}^3$ denotes the three-dimensional sphere embedded in R^4, $q = (q_v, q_4)$ includes vector $q_v \in R^3$ and scalar $q_4 \in R$ and satisfies the constraint $q_v^T q_v + q_4^2 = 1$, $\omega \in R^3$ represents the angular velocity, $J \in R^{3\times 3}$ denotes the constant symmetric positive-definite inertia matrix of the spacecraft, $u = [u_{\tau 1}, u_{\tau 2}, u_{\tau 3}]^T \in R^3$ denotes the control torque, $d \in R^3$ represents the external disturbances, and the operation $(\cdot)^\times \in R^{3\times 3}$ denotes a skew-symmetric matrix, that is

$$z^\times = \begin{bmatrix} 0 & -z_3 & z_2 \\ z_3 & 0 & -z_1 \\ -z_2 & z_1 & 0 \end{bmatrix}, \quad \forall z = [z_1, z_2, z_3] \in R^3. \quad (3)$$

Assumption 1 ([21,23]). *The desired angular velocity ω_d and its first derivative are bounded by $\|\omega_d\| \leq c_1$ and $\|\dot{\omega}_d\| \leq c_2$, respectively, where c_1 and c_2 are known positive constants.*

Assumption 2 ([8,33]). *Assume that the disturbance d is bounded by $\|d\| \leq l_g$ where l_g is a known positive constant.*

Assumption 3 ([29]). *The inertia matrix J is bounded by $\|J\| \leq J_M$, where J_M is a known positive constant.*

Property 1 ([29]). *The following properties hold for the skew-symmetric matrices $a^\times$ and $b^\times$ with $a, b \in R^3$*

$$a^\times b^\times = ba^T - a^T b I_3. \quad (4)$$

$$a^\times b = -b^\times a. \quad (5)$$

$$\|a^\times\| = \|a\|. \quad (6)$$

Property 2 ([29]). *The matrix $(C\omega_d)^\times J + J(C\omega_d)^\times$ is skew-symmetric matrix and C is the rotation matrix.*

2.2. Problem Statement

The desired attitude $q_d = (q_{dv}^T, q_{d4})^T \in R^3 \times R$ is defined by [8,32]

$$\begin{cases} \dot{q}_{dv} = \frac{1}{2}(q_{d4}I_3 + q_{dv}^\times)\omega_d, \\ \dot{q}_{d4} = -\frac{1}{2}q_{dv}^T\omega_d. \end{cases} \quad (7)$$

The relative attitude tracking error of the spacecraft is defined by $q_e = (e_v^T, e_4)^T \in \bar{S}^3$ where $e_v = [e_1, e_2, e_3] \in R^3$ and $e_4 \in R$. Then, the attitude tracking problem can be described as follows.

$$\begin{cases} \dot{e}_v = \frac{1}{2}(e_4 I_3 + e_v^\times)\omega_e, \\ \dot{e}_4 = -\frac{1}{2}e_v^T\omega_e. \end{cases} \quad (8)$$

$$J\dot{\omega}_e = -\omega^\times J\omega + J(\omega_e^\times C\omega_d - C\dot{\omega}_d) + \Gamma u + d. \quad (9)$$

$$\omega_e = \omega - C\omega_d. \quad (10)$$

where the diagonal matrix $\Gamma = diag(\gamma_1(t), \gamma_2(t), \gamma_3(t)) \in R^{3\times 3}$ denotes the actuator health condition and $\gamma_i(t)$ satisfies $\gamma_0 \leq \gamma_i(t) \leq 1$, $(i = 1,2,3)$ with a known positive constant γ_0. Clearly, $\gamma_i(t) = 1$ indicates the fault-free spacecraft and $\gamma_0 \leq \gamma_i(t) \leq 1$ denotes that the ith actuator partially loses its power [18,19]. The rotation matrix C is defined by $C = (e_4^2 - e_v^T e_v)I_3 + 2e_v e_v^T - 2e_4 e_v^\times$ where $\|C\| = 1$ and $\dot{C} = -\omega_e^\times C$ [8]. The error quaternion (e_v, e_4) satisfies $e_v^T e_v + e_4^2 = 1$.

We assume that exact attitude and velocity measurements are available and each actuator has a known maximum torque $u_{\tau i, max}$ satisfying

$$|u_{\tau i, max}| > J_M(c_1^2 + c_2) + l_g. \quad (11)$$

In this paper, the objective is to develop an adaptive hybrid fault-tolerant control law u subject to actuator constraints given by (11) to guarantee that the attitude tracking errors converge to the equilibrium point asymptotically without the unwinding phenomenon, which means $(0,0,0,\pm 1)^T$ is global stability for the rigid spacecraft in the presence of the actuator fault described by (8) and (9).

$$|u_{\tau i}| \leq u_{\tau i, max}. \quad (12)$$

3. Control Design

3.1. Hybrid System

Motivated by the work in [25,29], the following hysteresis-based hybrid function is introduced firstly to avoid the unwinding phenomenon.

$$\hbar \begin{cases} \dot{x} = M(x), & x \in D, \\ x^+ = N(x), & x \in E. \end{cases} \quad (13)$$

where the flow map $M : R^n \to R^n$ belongs to the flow set D, the jump map $N : R^n \to R^n$ belongs to the jump set E and x^+ represents the state value immediately after a jump [29].

Based on the hybrid system, we first introduce the following coordinate transformation S

$$S = \omega_e + \hbar\gamma^2 e_v. \quad (14)$$

where γ denotes the update law defined in (20) and the auxiliary variable $h \in \hbar = \{-1, 1\}$ satisfies $h^+ = -h$. The continuous set D and the jump set E are defined, respectively, as follows.

$$D = \left\{ x \in S^3 \times R^3 \times \hbar : he_4 > -\eta \right\}. \tag{15}$$

$$E = \left\{ x \in S^3 \times R^3 \times \hbar : he_4 \leq -\eta \right\}. \tag{16}$$

where $x = \{q_e, \omega_e, h\}$, $\eta \in (0, 1)$ indicates the hysteresis gap.

Remark 1. *It is worth noting that h is chosen to change the desired rotation direction to push q_e to either $(0, 0, 0, 1)^T$ or $(0, 0, 0, -1)^T$. Thus, the desired rotation direction changes only when there is a significant benefit in switching it, where "significant" is defined precisely by the selection of η. The hysteresis width η manages a trade-off between robustness to disturbance and a small amount of hysteresis-induced inefficiency [25].*

3.2. Controller Formulation

The hybrid saturation and fault-tolerant attitude control (HSFC) is proposed as:

$$u = u_1 + u_2. \tag{17}$$

where

$$u_1 = -k \frac{S}{(|S_i| + \gamma^2 \delta)} + (C\omega_d)^\times JC\omega_d + JC\dot{\omega}_d. \tag{18}$$

$$u_2 = -\left(\frac{1 - \gamma_0}{\gamma_0} \|u_1\| \right) sign(\omega_e). \tag{19}$$

$$\dot{\gamma} = \frac{\alpha \gamma}{1 + 2\alpha k_1(1 - he_4)} \left(k \sum_{i=1}^{3} \left(\frac{he_{vi}\omega_{ei}}{(|S_i| + \gamma^2 \delta)} - \frac{|\omega_{ei}|(1 + \delta)}{|\omega_{ei}| + \gamma^2(1 + \delta)} \right) - \frac{1}{2} k_1 h e_v^T S \right). \tag{20}$$

where $sign(\cdot)$ denotes the sign function, k, k_1, α, δ are positive constants and $k > l_g$.

Remark 2. *It should be pointed out that the equilibrium point $(0, 0, 0, \pm 1)^T$ represents the same physical attitude for rigid spacecraft formulated by quaternion. When this double covering is neglected, the traditional controller can induce the notion called "unwinding", which leads to the spacecraft making an unnecessarily full rotation [25] and consuming more unnecessary energy. The proposed HSFC is designed to tackle the unwinding phenomenon.*

Utilizing the facts that $\|C\| = 1$, $\|e_4 I_3 + e_v^\times\| = 1$, $|e_i| \leq 1$ and $h^2 = 1$, the control torque u given by (17) can be upper bounded by

$$|u_{\tau i}| \leq \frac{1}{\gamma_0} \left(k + J_M \left(c_1^2 + c_2 \right) \right), \quad u = [u_{\tau 1}, u_{\tau 2}, u_{\tau 3}]^T. \tag{21}$$

Rewriting Γu as $u_1 - (I_3 - \Gamma)u_1 + \Gamma u_2$ and utilizing the fact $\omega = \omega_e + C\omega_d$, we have

$$\begin{aligned} J\dot{\omega}_e = & -\omega_e^\times J\omega_e - \omega_e^\times JC\omega_d - (C\omega_d)^\times J\omega_e - (C\omega_d)^\times JC\omega_d \\ & + J(\omega_e^\times C\omega_d - C\dot{\omega}_d) + u_1 - (I_3 - \Gamma)u_1 + \Gamma u_2 + d. \end{aligned} \tag{22}$$

4. Stability Analysis

Now, Theorem 1 of the main result of this paper is stated as follows.

Theorem 1. *Considering the rigid spacecraft described as (8) and (9), the developed approach defined by (17)–(19) ensures the attitude tracking errors globally converge to the equilibrium point asymptotically.*

Proof. The proof includes the following two consecutive main steps. First, when $x \in D$, all the states are continuous and h remains unchanged such that $\dot{h} = 0$; we prove that system states are stable in set D by using LaSalle's invariance principle for the hybrid system. Second, when $x \in E$, the jump only occurs with the variable h and the other system states are still continuous; we prove that system states are stable in set E.

Step 1. The following positive-definite Lyapunov function candidate is proposed.

$$V = \gamma^2 \frac{k_1}{2}\left[(1-he_4)^2 + e_v^T e_v\right] + \frac{1}{2}\omega_e^T J \omega_e + \frac{\gamma^2}{2\alpha}. \quad (23)$$

Note that the following equality holds with the fact $e_v^T e_v + e_4^2 = 1$.

$$(1-he_4)^2 + e_v^T e_v = 1 - 2he_4 + e_4^2 + e_v^T e_v = 2 - 2he_4 = 2(1-he_4). \quad (24)$$

In light of (24), we can rewrite (23) as

$$V = \gamma^2 k_1 (1 - he_4) + \frac{1}{2}\omega_e^T J \omega_e + \frac{\gamma^2}{2\alpha}. \quad (25)$$

The time derivative of V along (22) takes

$$\dot{V} = -\gamma^2 k_1 h \dot{e}_4 + \omega_e^T J \dot{\omega}_e + \dot{\gamma}\gamma\left(\frac{1}{\alpha} + 2k_1(1-he_4)\right). \quad (26)$$

By virtue of the fact $\dot{e}_4 = -\frac{1}{2}e_v^T \omega_e$, (26) can be further formulated as

$$\dot{V} = \frac{1}{2}\gamma^2 k_1 h e_v^T \omega_e + \omega_e^T J \dot{\omega}_e + \dot{\gamma}\gamma\left(\frac{1}{\alpha} + 2k_1(1-he_4)\right). \quad (27)$$

When $x \in D$, substituting $J\dot{\omega}_e$ from (22) into (27), it follows that

$$\begin{aligned}\dot{V} &= \frac{1}{2}\gamma^2 k_1 h e_v^T \omega_e + \dot{\gamma}\gamma\left(\frac{1}{\alpha} + 2k_1(1-he_4)\right) \\ &+ \omega_e^T \begin{pmatrix} -\omega_e^\times J \omega_e - \omega_e^\times JC\omega_d - (C\omega_d)^\times J\omega_e - (C\omega_d)^\times JC\omega_d \\ +J(\omega_e^\times C\omega_d - C\dot{\omega}_d) + u_1 - (I_3 - \Gamma)u_1 + \Gamma u_2 + d \end{pmatrix}.\end{aligned} \quad (28)$$

Using Properties 1 and 2, this yields

$$\begin{aligned}&\omega_e^T \omega_e^\times = 0, \\ &\omega_e^\times C\omega_d = -(C\omega_d)^\times \omega_e, \\ &\omega_e^T \left((C\omega_d)^\times J + J(C\omega_d)^\times\right)\omega_e = 0.\end{aligned} \quad (29)$$

Upon utilizing the above facts, Equation (28) yields

$$\begin{aligned}\dot{V} &= \frac{1}{2}\gamma^2 k_1 h e_v^T \omega_e + \dot{\gamma}\gamma\left(\frac{1}{\alpha} + 2k_1(1-he_4)\right) \\ &+ \omega_e^T \left(u_1 + d - (C\omega_d)^\times JC\omega_d - JC\dot{\omega}_d\right) + \omega_e^T(-(I_3 - \Gamma)u_1 + \Gamma u_2).\end{aligned} \quad (30)$$

Upon substituting the controller (18) into (30) and applying the fact that $\omega_e = S - h\gamma^2 e_v$, we obtain

$$\begin{aligned}\dot{V} &= \frac{1}{2}\gamma^2 k_1 h e_v^T (S - h\gamma^2 e_v) + \dot{\gamma}\gamma\left(\frac{1}{\alpha} + 2k_1(1-he_4)\right) \\ &+ \omega_e^T d + \omega_e^T(-(I_3 - \Gamma)u_1 + \Gamma u_2) - k\sum_{i=1}^{3}\frac{\omega_{ei}S_i}{(|S_i| + \gamma^2 \delta)}.\end{aligned} \quad (31)$$

In light of (14), we can rewrite (31) as

$$\dot{V} = \tfrac{1}{2}\gamma^2 k_1 h e_v^T (S - h\gamma^2 e_v) + \dot{\gamma}\gamma \left(\tfrac{1}{\alpha} + 2k_1(1-he_4)\right) \\ + \omega_e^T d + \omega_e^T(-(I_3 - \Gamma)u_1 + \Gamma u_2) - k \sum_{i=1}^{3} \tfrac{\omega_{ei}^2 + h\gamma^2 e_{vi}\omega_{ei}}{(|S_i|+\gamma^2\delta)}. \tag{32}$$

By virtue of the triangle inequality, we have

$$\sum_{i=1}^{3} \tfrac{\omega_{ei}^2}{(|S_i|+\gamma^2\delta)} = \sum_{i=1}^{3} \tfrac{\omega_{ei}^2}{(|\omega_{ei}+h\gamma^2 e_{vi}|+\gamma^2\delta)}, \\ \sum_{i=1}^{3} \tfrac{\omega_{ei}^2}{(|\omega_{ei}+h\gamma^2 e_{vi}|+\gamma^2\delta)} \geq \sum_{i=1}^{3} \tfrac{\omega_{ei}^2}{(|\omega_{ei}|+\gamma^2(1+\delta))}, \\ \sum_{i=1}^{3} \tfrac{\omega_{ei}^2}{(|\omega_{ei}|+\gamma^2(1+\delta))} = \sum_{i=1}^{3} \left(|\omega_{ei}| - \tfrac{|\omega_{ei}|\gamma^2(1+\delta)}{|\omega_{ei}|+\gamma^2(1+\delta)}\right). \tag{33}$$

Upon applying (33) and the facts that $\sum_{i=1}^{3} |\omega_{ei}| \geq \|\omega_e\|$ and $\|d\| \leq l_g$ to (32), we obtain

$$\dot{V} \leq \tfrac{1}{2}\gamma^2 k_1 h e_v^T(S - h\gamma^2 e_v) + \dot{\gamma}\gamma\left(\tfrac{1}{\alpha} + 2k_1(1-he_4)\right) \\ + \omega_e^T d + \omega_e^T(-(I-\Gamma)u_1 + \Gamma u_2) \\ - k\sum_{i=1}^{3}\left(|\omega_{ei}| - \tfrac{|\omega_{ei}|\gamma^2(1+\delta)}{|\omega_{ei}|+\gamma^2(1+\delta)}\right) - k\sum_{i=1}^{3}\tfrac{h\gamma^2 e_{vi}\omega_{ei}}{(|S_i|+\gamma^2\delta)}. \tag{34}$$

After substituting the update law (20) into (34), we obtain

$$\dot{V} \leq -\tfrac{1}{2}\gamma^4 k_1 h^2 e_v^T e_v + \|\omega_e\| l_g - k\|\omega_e\| + \omega_e^T(-(I_3 - \Gamma)u_1 + \Gamma u_2). \tag{35}$$

Recalling the facts that $\sum_{i=1}^{3} |\omega_{ei}| \geq \|\omega_e\|$, $\gamma_0 \leq \gamma_i$ and $\|I_3 - \Gamma\| = \lambda_M(I_3 - \Gamma) \leq 1 - \gamma_0$, we have

$$\omega_e^T \Gamma u_2 = \omega_e^T \Gamma\left(-\left(\tfrac{1-\gamma_0}{\gamma_0}\|u_1\|\right)sign(\omega_e)\right) \leq -\lambda_{\min}(\Gamma)\left(\tfrac{1-\gamma_0}{\gamma_0}\|u_1\|\right)\sum_{i=1}^{3}|\omega_{ei}| \\ \leq -\gamma_0\left(\tfrac{1-\gamma_0}{\gamma_0}\|u_1\|\right)\sum_{i=1}^{3}|\omega_{ei}| = -(1-\gamma_0)\|u_1\|\sum_{i=1}^{3}|\omega_{ei}| \\ \leq -(1-\gamma_0)\|u_1\|\|\omega_e\|. \tag{36}$$

Substituting (19) and (36) into (35) yields

$$\dot{V} \leq -c\|\omega_e\| - \tfrac{1}{2}\gamma^4 k_1 h^2 e_v^T e_v + \|\omega_e\|(1-\gamma_0)\|u_1\| - (1-\gamma_0)\|u_1\|\|\omega_e\|. \tag{37}$$

In light of $h^2 = 1$, (37) can be rewritten as

$$\dot{V} \leq -c\|\omega_e\| - \tfrac{1}{2}\gamma^4 k_1 e_v^T e_v. \tag{38}$$

where

$$c = k - l_g > 0. \tag{39}$$

When $x \in E$, the jump occurs in V and we have

$$V(x^+) - V(x) = 2\gamma^2 k_1 h e_4. \tag{40}$$

In view of (16), we obtain

$$V(x^+) - V(x) \leq -2\gamma^2 k_1 \eta < 0. \tag{41}$$

It is clear from (38) that $\dot{V} \leq 0$ where $c > 0$, $\gamma^4 > 0$ and $k_1 > 0$, when $x \in D$. Moreover, $\dot{V} = 0$ implies that $e_v = 0$ and $\omega_e(t) = 0$. Otherwise, when $x \in E$, we can conclude that $\dot{V} \leq 0$ from (41). Hence, by applying LaSalle's invariance theorem [34] and theorem 7.6 from [35], we can conclude that $\lim_{t\to\infty} e_v(t) = 0$ and $\lim_{t\to\infty} \omega_e(t) = 0$.

Step 2. When $x \in E$, no jump occurs. Thus, we have

$$V(x^+) - V(x) = 0. \tag{42}$$

Actually, the set $\{x \in E : V(x^+) - V(x) = 0\}$ is empty. Using Theorem 4.7 in [35], the tracking errors converge to the largest invariant set $\Psi = \{(\omega_e, q_e, h) | \dot{V} = 0, he_4 \geq -\eta\}$. By virtue of (38), it is clear that $\dot{V} = 0$ means that $e_v = 0$ and $\omega_e(t) = 0$. □

Remark 3. *Comparing with our recent work in [5,32,33], the proposed control not only can guarantee the control torque of the actuator can be upper bounded a priori by selecting the controller parameters but also can compensate the partial failure of the actuator. This is in contrast to the work of [29,30] who only tackle the attitude tracking problem for the fault-free spacecraft system.*

Remark 4. *The saturation vector $W(\omega_e) = [w(\omega_{e_1}), w(\omega_{e_2}), w(\omega_{e_3})]^T$ is used to eliminate chattering caused by the discontinuous vector function $sign(\omega_e)$ in controller (19) and $w(\omega_{e_i})$ is given by*

$$w(\omega_{e_i}) = \begin{cases} \frac{\omega_{e_i}}{|\omega_{e_i}|} & |\omega_{e_i}| > \bar{\varepsilon}, \\ \frac{\omega_{e_i}}{\bar{\varepsilon}} & |\omega_{e_i}| \leq \bar{\varepsilon}. \end{cases} \tag{43}$$

where $\bar{\varepsilon}$ is a small positive constant.

Remark 5. *To avoid control torque over the real actuator maximum output (that is actuator saturation) in advance, the parameters k and γ_0 are chosen to constrain the control amplitude, and k and γ_0 satisfy $k > l_g$ and $\gamma_0 \leq \gamma_i(t) \leq 1$, respectively. Moreover, γ_0 in Equation (19) is designed to compensate the partial loss of actuator effectiveness faults. For a healthy actuator, γ_0 is chosen as $\gamma_0 = 1$, while γ_0 is selected as $\gamma_0 \leq \gamma_i(t) \leq 1$ for the actuator partial loss effectiveness to compensate the fault. Finally, to avoid the unwinding phenomenon, $h = \{1, -1\}$ is chosen to change the desired rotation direction to push q_e to either $(0,0,0,1)^T$ or $(0,0,0,-1)^T$. In addition, a large value of $0 < \delta < 1$ will decrease the convergence rate.*

5. Simulation

The simulations are performed on the spacecraft used in [2] to illustrate the effectiveness and the improved performance of the proposed HSFC. It should be pointed out that the parameters used in the simulation except actuator failure are completely the same as [2]. The inertia matrix is $J = [22\ 1.2\ 0.9; 1.2\ 19\ 1.4; 0.9\ 1.4\ 18]$ kg·m^2. The desired angular velocity is selected as $\omega_d(t) = 0.05\ [\sin(\pi t/100), \sin(2\pi t/100), \sin(3\pi t/100)]$ rad/sec, the desired attitude is generated by (7) and the initial desired attitude is chosen as $q_d(0) = [0, 0, 0, 1]^T$. The initial update law is chosen as $\gamma(0) = 2.5$. The initial attitude and angular velocity of the spacecraft are $q(0) = [0.3, -0.2, -0.3, 0.8832]^T$ and $\omega(0) = [0.06, -0.04, 0.05]^T$ rad/sec, respectively. The external disturbance is chosen as $d(t) = [0.1\sin(t), 0.2\sin(1.2t), 0.3\sin(1.5t)]$ N·m.

In light of the above system parameters and utilizing Assumptions 1–3, we obtain

$$J_M = 22.8\ \text{kg}\cdot\text{m}^2,\ l_g = 0.3\ \text{N}\cdot\text{m},\ c_1 = 5\times10^{-2},\ c_2 = 4.71\times10^{-3}. \tag{44}$$

5.1. Verification of the Effectiveness of HSFC with Fault Compensation

The comparison is performed on both HSFC and HSFC without fault compensation term u_2 for the actuator faulted spacecraft to verify the fault-tolerant property of the proposed HSFC. It is assumed that the actuator failure matrix is chosen as $\Gamma = diag(0.5 +$

$0.01\sin(10t), 0.5 + 0.02\cos(20t), 0.5 + 0.03\sin(30t))$ and $\gamma_0 = 0.2$. The other parameters of the proposed HSFC are chosen as $k = 5, k_1 = 5, k_2 = 5, \alpha = 0.01, \bar{\varepsilon} = 0.005$ and $\delta = 0.05$.

The maximum torque of the actuator in practical system is assumed to be $|u_{\tau i,\max}| = 10\,\text{N}\cdot\text{m}$. According to Equations (21) and (44), the upper bound of the control torque is $8.6\,\text{N}\cdot\text{m}$ and satisfies $|u_{\tau i}| \leq 8.6\,\text{N}\cdot\text{m} \leq |u_{\tau i,\max}| = 10\,\text{N}\cdot\text{m}$, which means that the proposed HSFC can be an anti-saturated controller, due to the maximum actual control torque being constrained to $10\,\text{N}\cdot\text{m}$.

The simulation results of the HSFC without fault compensation term u_2 for the actuator faulted spacecraft are shown in Figures 1–3, while those of the HSFC with fault compensation term u_2 are illustrated in Figures 4–6. Clearly, the HSFC with fault compensation term u_2 converges to the equilibrium point fast due to the fault-tolerant property, as we see in Figures 4 and 5, while the HSFC without the fault compensation term u_2 takes more time to complete its tracking, as we see in Figures 1 and 2.

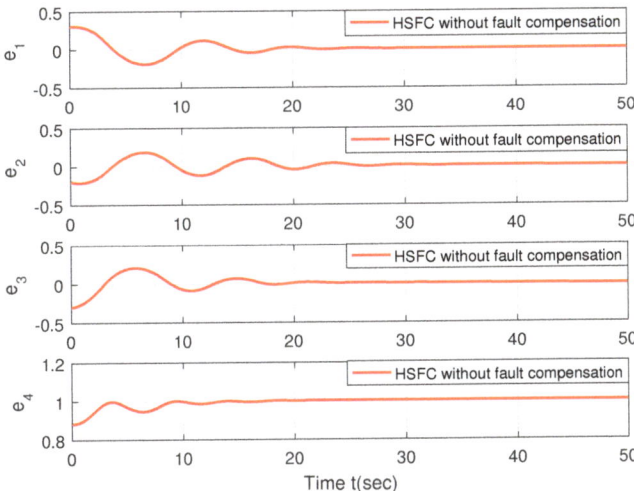

Figure 1. Attitude tracking errors.

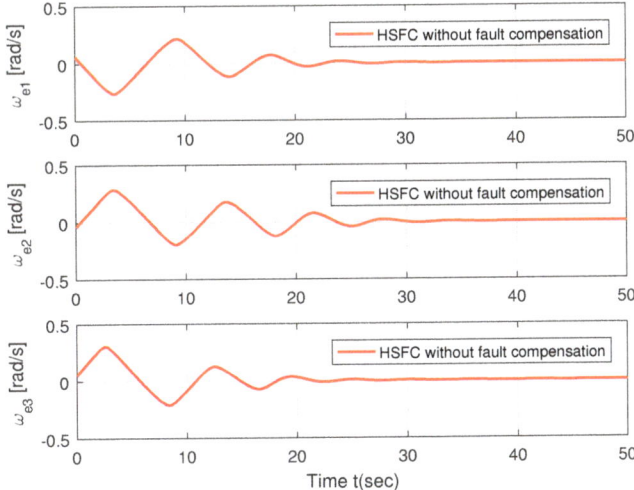

Figure 2. Angular velocity tracking errors.

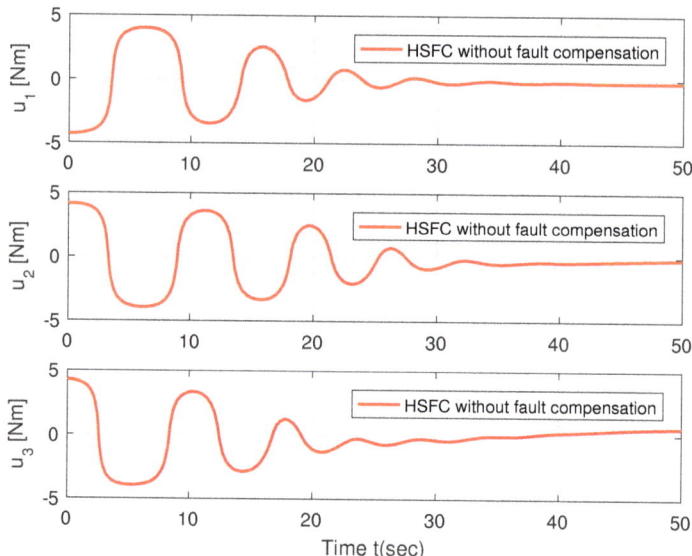

Figure 3. Control torque.

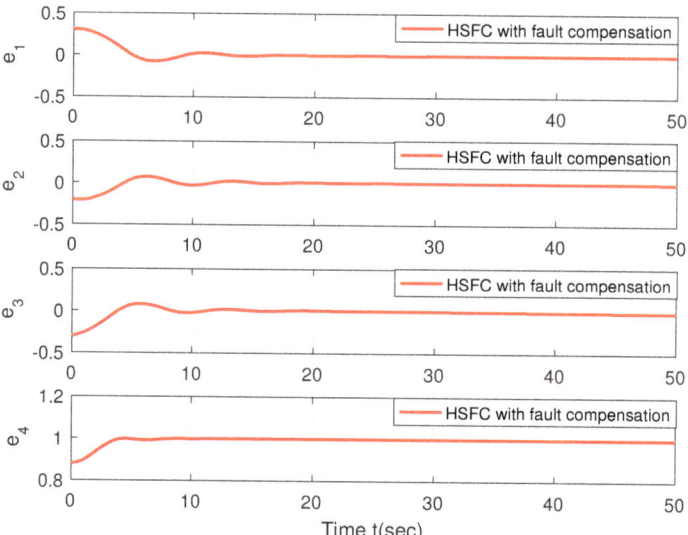

Figure 4. Attitude tracking errors.

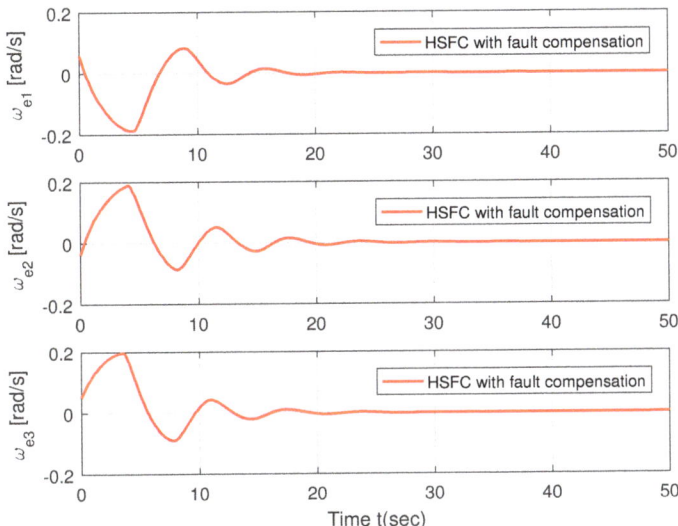

Figure 5. Angular velocity tracking errors.

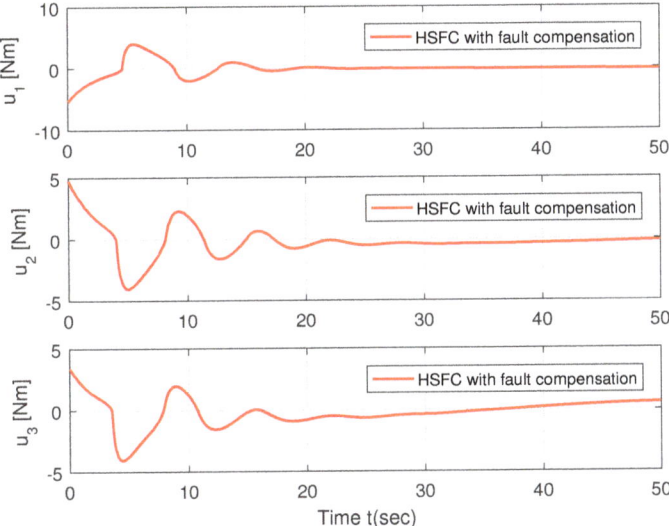

Figure 6. Control torque.

5.2. Comparisons with the AFNTSMC and SHOPD+

Firstly, a comparison with the AFNTSMC in [2] is performed to show the anti-unwinding performance of the proposed HSFC. Because the AFNTSMC does not consider actuator failure, the comparison is conducted on the fault-free spacecraft. Thus, the matrix $\Gamma = diag(1.0, 1.0, 1.0)$ is chosen to describe the healthy actuator and $\gamma_0 = 1$. The AFNTSMC is formulated as follows.

$$u = -\left(\tau + u_{adp}(t)\right)S(t) - \beta_0 sig^{\chi_0}(S). \tag{45}$$

$$S = \omega_e + \bar{k}_2 e_v + \bar{k}_3 S_{au}. \tag{46}$$

$$S_{aui} = \begin{cases} e_i^\nu, & \text{if } \bar{S}_i = 0 \text{ or } \bar{S}_i \neq 0, |e_i| \geq \varepsilon, \\ \iota_1 e_i + \iota_2 sign(e_i) e_i^2, & \text{if } \bar{S}_i \neq 0, |e_i| < \varepsilon. \end{cases} \quad (47)$$

and $\iota_1 = (2-\nu)\varepsilon^{\nu-1}$, $\iota_2 = (\nu-1)\varepsilon^{\nu-2}$.

$$\bar{S}_i = \omega_{ei} + \bar{k}_2 e_i + \bar{k}_3 e_i^\nu. \quad (48)$$

$$u_{adp} = diag(\hat{\chi}_i). \quad (49)$$

$$\hat{\chi}_i = \frac{1}{2}\varepsilon_{5i}^{-2}\hat{\psi}_i + \frac{1}{2}\varepsilon_{6i}^{-2}\hat{\phi}_i\|\xi\|^2. \quad (50)$$

$$\dot{\hat{\psi}}_i(t) = -\varepsilon_{3i}\hat{\psi}_i(t) + \frac{1}{2}n_{1i}\varepsilon_{5i}^{-2}|S_i(t)|^2. \quad (51)$$

$$\dot{\hat{\phi}}_i(t) = -\varepsilon_{4i}\hat{\phi}_i(t) + \frac{1}{2}n_{2i}\varepsilon_{6i}^{-2}|S_i(t)|^2\|\xi\|^2. \quad (52)$$

where τ and β_0 are diagonal constant matrices, and $\tau_i, \beta_{0i}, i = 1, 2, 3, k_2$ and k_3 are positive constants, ε is a small positive constant, $0 < \chi_0 < 1, \nu_1, \nu_2$ are positive odd integers and satisfy $0 < \nu = \frac{\nu_1}{\nu_2} < 1$, and $\|\xi\| = \max\{\|\omega\|^2, \|\omega\|\}$.

The initial conditions are changed to $q(0) = [0.3, -0.2, -0.3, -0.8832]^T$ and $\omega(0) = [0.06, -0.04, 0.05]^T$ rad/sec to verify the anti-unwinding performance. The following parameters of the AFNTSMC are chosen the same as [2]: $\varepsilon_{3i} = \varepsilon_{4i} = 0.35$, $\varepsilon_{5i} = \varepsilon_{6i} = 0.16, \bar{k}_2 = \bar{k}_3 = 1, \chi_0 = 0.5, \varepsilon = 0.001, \tau_i = 20, \nu_2 = 5, \nu_1 = 3, n_{1i} = n_{2i} = 6$, $\beta_{0i} = 10$ and $\hat{\psi}_i(0) = \hat{\phi}_i(0) = 0.1$. The parameters of the proposed HSFC are selected as $k = 5, k_1 = 5, k_2 = 5, \alpha = 0.1, \bar{\varepsilon} = 0.005$ and $\delta = 0.01$.

The comparison results are shown in Figures 7–10. From the comparison of Figure 7, it is clearly seen that the proposed HSFC can guarantee the attitude tracking errors converge to the equilibrium point $(0, 0, 0, -1)$ instead of $(0, 0, 0, 1)$, which means that the unwinding phenomenon is tackled in comparison with the AFNTSMC. It is important to note that the unwinding property of the proposed HSFC has the benefit of decreasing excessive energy consumption compared to AFNTSMC, as we see in Figure 10.

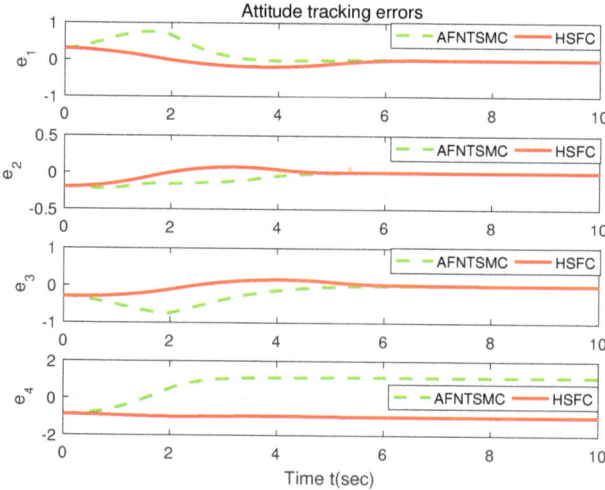

Figure 7. Comparison of attitude tracking errors.

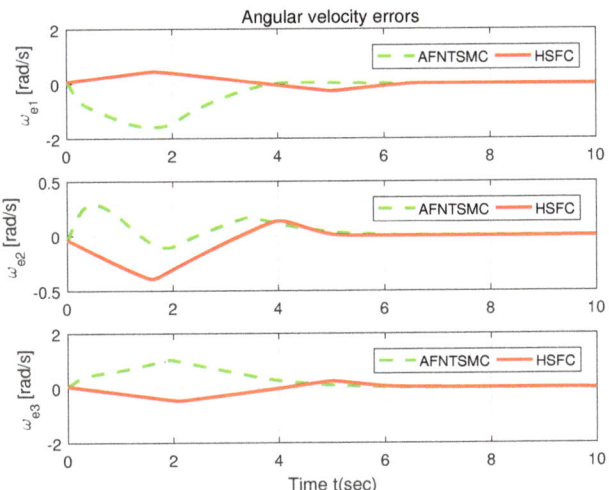

Figure 8. Comparison of angular velocity tracking errors.

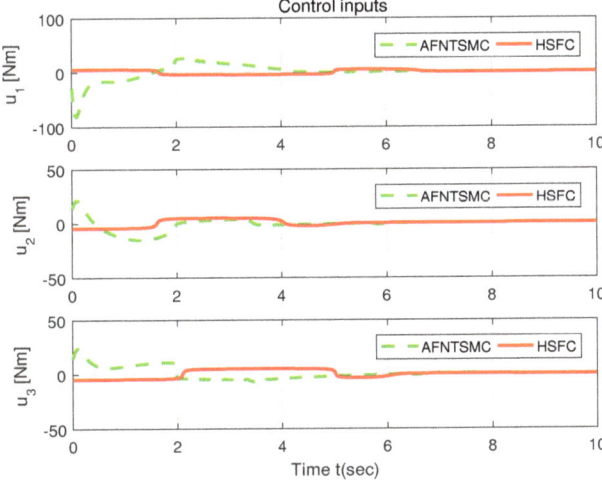

Figure 9. Comparison of control torque.

Secondly, a comparison with SHOPD+ in [29] is also illustrated to show the improved performance of the proposed HSFC. Both HSFC and SHOPD+ are anti-unwinding controllers. The SHOPD+ is given as follows

$$u = -k_4 h e_v - k_5 (e_4 I_3 + e_v^\times)^T S_a + (C\omega_d)^\times JC\omega_d + JC\dot{\omega}_d, \qquad (53)$$

$$S_a(\nu_i) = \begin{cases} sign(\nu_i), & |\nu_i| \geq 1 \\ \nu_i, & |\nu_i| < 1 \end{cases}, \qquad (54)$$

$$\begin{cases} \nu = q_c + Be_v \\ \dot{q}_c = -A\nu \end{cases}. \qquad (55)$$

The parameters of the SHOPD+ are selected as: $k_4 = 30$ and $k_5 = 8$, $A = diag(1,1,1)$ and $B = diag(3,3,3)$.

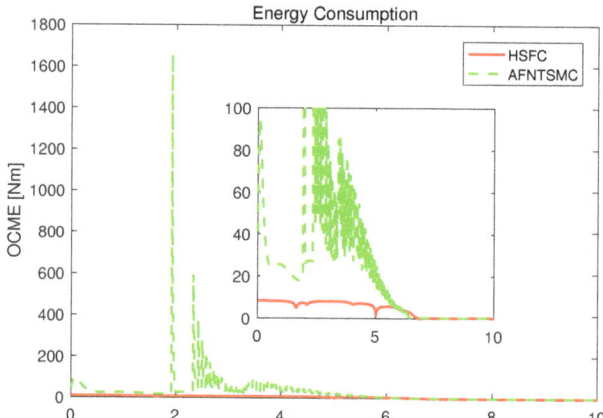

Figure 10. Comparison of energy consumption.

The comparison results are demonstrated in Figures 11–13. Obviously, HSFC and SHOPD+ can completely track their desired attitude and angular velocity within the allowable torques. Compared with SHOPD+, the proposed HSFC can achieve a fast transient over the SHOPD+ due to the fault compensation ability of HSFC, as we see in Figures 11 and 12. Moreover, the proposed HSFC has the benefit of decreasing excessive control torque compared to SHOPD+, as we see in Figure 13.

Based on the above simulation results, one can conclude that the designed HSFC can tackle the actuator saturation and partial loss failure problem of rigid spacecraft subject to external disturbances. Furthermore, in contrast to AFNTSMC in [2], the proposed HSFC also can overcome the unwinding phenomenon of rigid spacecraft. Compared with SHOPD+ in [29], the proposed HSFC can obtain a fast transient and compensate the actuator failure within the allowable torques of spacecraft.

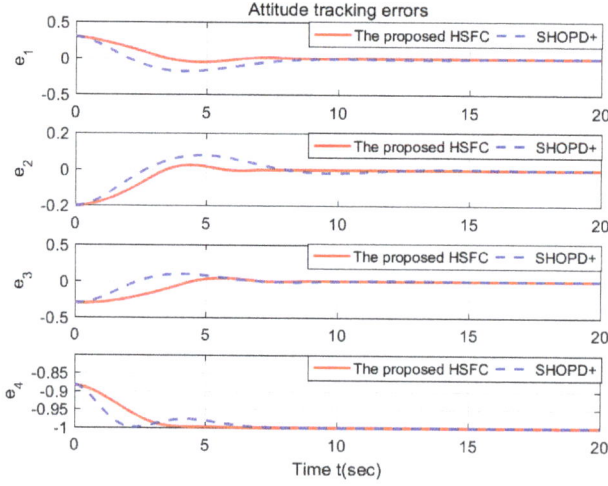

Figure 11. Comparison of attitude tracking errors.

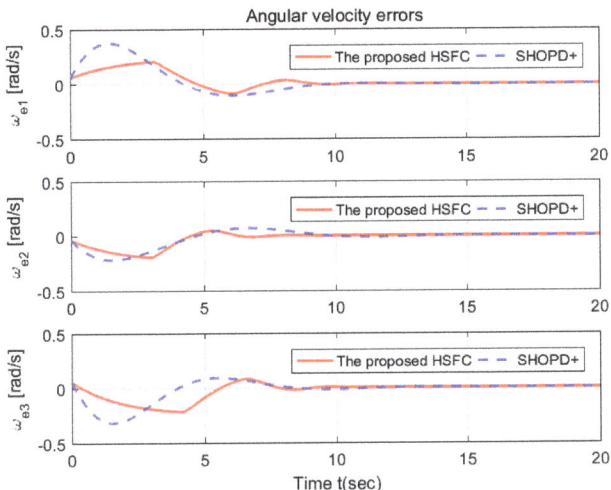

Figure 12. Comparison of angular velocity tracking errors.

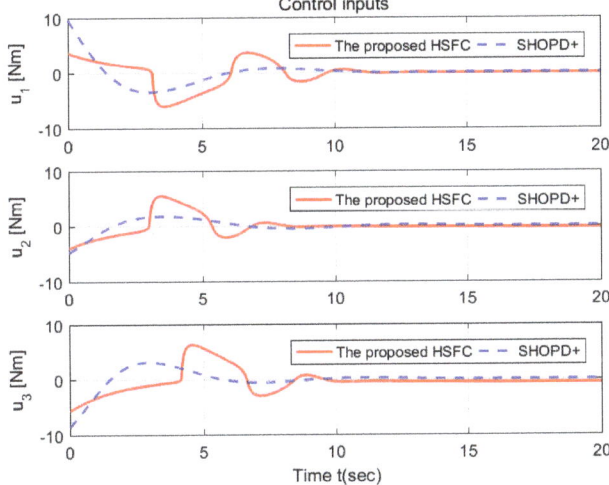

Figure 13. Comparison of control torque.

6. Conclusions

In this paper, a robust hybrid saturation and fault-tolerant control has been proposed for rigid spacecraft subject to external disturbances and actuator partial loss failure. The proposed HSFC can avoid actuator saturation and partial loss failure by selecting the control gains in advance, which implies that degraded performance of the actuator or unpredictable attitude tracking can be completely eliminated. Lyapunov's method is borrowed to prove the global asymptotic stability. The main features of the proposed HSFC include actuator saturation, fault-tolerance and robustness. Simulations verify the effectiveness and improved performance of the proposed control.

Author Contributions: Conceptualization, J.M. and Z.W.; methodology, Z.W.; software, Z.W.; validation, J.M., Z.W. and C.W.; formal analysis, C.W.; investigation, C.W.; resources, J.M.; data curation, J.M.; writing—original draft preparation, Z.W.; writing—review and editing, Z.W.; visualization, C.W.; supervision, C.W.; project administration, Z.W.; funding acquisition, Z.W. All authors have read and agreed to the published version of the manuscript.

Funding: This work was supported by the Education Department of Shaanxi Province with Grant/Award Numbers 21JK0675 and the Natural Science Foundation of Shaanxi Province with Grant/Award Numbers 2023-JC-QN-0738.

Institutional Review Board Statement: Not applicable.

Informed Consent Statement: Not applicable.

Data Availability Statement: The data that support the findings of this study are available from the corresponding author upon reasonable request.

Acknowledgments: The authors would like to thank the anonymous reviewers, Associate Editor and Editor for their valuable comments, which are helpful to improve the quality of this article.

Conflicts of Interest: The authors declare no conflict of interest.

References

1. Egeland, O.; Godhavn, J.M. Passivity-based adaptive attitude control of a rigid spacecraft. *IEEE Trans. Autom. Control* **1994**, *39*, 842–846. [CrossRef]
2. Lu, K.F.; Xia, Y.Q. Adaptive attitude tracking control for rigid spacecraft with finite-time convergence. *Automatica* **2013**, *49*, 3591–3599. [CrossRef]
3. Wang, Z.; Su, Y.X.; Zhang, L.Y. A new nonsingular terminal sliding mode control for rigid spacecraft attitude tracking. *J. Dyn Syst. Meas. Control* **2018**, *140*, 051006. [CrossRef]
4. Wang, Z.; Su, Y.X.; Zhang, L.Y. Fixed-time attitude tracking control for rigid spacecraft. *IET Control Theory Appl.* **2020**, *14*, 790–799. [CrossRef]
5. Wang, Z.; Ma, J.; Hu, L.M. Predefined-time guaranteed performance attitude takeover control for non-cooperative spacecraft with uncertainties. *Int. J. Robust Nonlinear Control* **2023**, *33*, 7488–7509. [CrossRef]
6. Dixon, W.E. Adaptive regulation of amplitude limited robot manipulators with uncertain kinematics and dynamics. *IEEE Trans Autom. Control* **2007**, *52*, 488–493. [CrossRef]
7. Perez-Arancibia, N.O.; Tsao, T.C.; Gibson, J.S. Saturation-induced instability and its avoidance in adaptive control of hard disk drives. *IEEE Trans. Control Syst. Technol* **2010**, *18*, 368–382. [CrossRef]
8. Boskovic, J.D.; li, S.M.; Mehra, R.K. Robust tracking control design for spacecraft under control input saturation. *J. Guid. Control Dyn.* **2004**, *27*, 627–633. [CrossRef]
9. Ali, I.; Radice, G.H.; Kim, J. Backstepping control design with actuator torque bound for spacecraft attitude maneuver. *J. Guid. Control Dyn.* **2010**, *33*, 254–259. [CrossRef]
10. de Ruiter, A.H.J.; Zheng, C.H. Adaptive spacecraft attitude tracking control with actuator saturation. *J. Guid. Control Dyn.* **2010**, *33*, 1692–1696. [CrossRef]
11. Su, Y.X.; Zheng, C.H. Globally asymptotic stabilization of spacecraft with simple saturated proportional-derivative control. *J. Guid. Control Dyn.* **2011**, *34*, 1932–1935. [CrossRef]
12. Su, Y.X.; Zheng, C.H. Velocity-free saturated PD controller for asymptotic stabilization of spacecraft. *Aerosp. Sci. Technol.* **2014**, *39*, 6–12. [CrossRef]
13. Su, Y.X.; Zheng, C.H. Unified saturated proportional derivative control framework for asymptotic stabilisation of spacecraft. *IET Control Theory Appl.* **2016**, *10*, 772–779. [CrossRef]
14. Su, Y.X.; Zheng, C.H. Single saturated PD control for asymptotic attitude stabilisation of spacecraft. *IET Control Theory Appl.* **2020**, *14*, 3338–3343. [CrossRef]
15. Xia, Y.Q.; Su, Y.X. Saturated output feedback control for finite-time attitude stabilization of spacecraft. *Proc. Inst. Mech. Eng. Part C J. Mech. Eng. Sci.* **2020**, *234*, 4557–4571. [CrossRef]
16. Yin, S.; Xiao, B.; Ding, S.X.; Zhou, D.H. A review on recent development of spacecraft attitude fault tolerant control system. *IEEE Trans. Ind. Electron.* **2016**, *63*, 3311–3320. [CrossRef]
17. Wang, Z.; Shan, J.J. Fixed-time consensus for uncertain multi-agent systems with actuator faults. *J. Frankl. Inst.* **2020**, *357*, 1199–1220. [CrossRef]
18. Cai, W.C.; Liao, X.H.; Song, D.Y. Indirect robust adaptive fault-tolerant control for attitude tracking of spacecraft. *J. Guid. Control Dyn.* **2008**, *31*, 1456–1463. [CrossRef]
19. Bustan, D.; Sani, S.H.; Pariz, N. Adaptive fault-tolerant spacecraft attitude control design with transient response control. *IEEE/ASME Trans. Mechatronics* **2014**, *19*, 1404–1411.
20. Shen, Q.; Wang, D.W.; Zhu, S.Q.; Poh, E.K. Inertia-free fault-tolerant spacecraft attitude tracking using control allocation. *Automatica* **2015**, *62*, 114–121. [CrossRef]
21. Wang, Z.; Su, Y.X.; Zhang, L.Y. Fixed-time fault-tolerant attitude tracking control for rigid spacecraft. *J. Dyn. Syst. Meas. Control* **2020**, *142*, 024502. [CrossRef]
22. Sun, R.; Shan, A.D.; Zhang, C.X.; Wu, J.; Jia, Q.X. Quantized fault-tolerant control for attitude stabilization with fixed-time disturbance observer. *J. Guid. Control Dyn.* **2021**, *44*, 449–455. [CrossRef]

3. Shen, Q.; Wang, D.W.; Zhu, S.Q.; Poh, E.K. Fault-tolerant optimal spacecraft attitude maneuver: An incremental model approach. *J. Guid. Control Dyn.* **2022**, *62*, 114–121.
4. Bhat, S.P.; Bernstein, D.S. A topological obstruction to continuous global stabilization of rotational motion and the unwinding phenomenon. *Syst. Control Lett.* **2012**, *61*, 595–601. [CrossRef]
5. Mayhew, C.G.; Sanfelice, R.G.; Teel, A.R. Quaternion-based hybrid control for robust global attitude tracking. *IEEE Trans. Autom. Control* **2011**, *56*, 2555–2566. [CrossRef]
6. Lee, T. Global Exponential Attitude Tracking Controls on SO(3). *IEEE Trans. Autom. Control* **2015**, *60*, 2837–2842. [CrossRef]
7. Schlanbusch, R.; Grotli, E.I. Hybrid certainty equivalence control of rigid bodies with quaternion measurements. *IEEE Trans. Autom. Control* **2015**, *60*, 2512–2517. [CrossRef]
8. Gui, H.C.; Vukovich, G. Global finite-time attitude tracking via quaternion feedback. *Syst. Control Lett.* **2016**, *97*, 176–183. [CrossRef]
9. Xia, Y.Q.; Su, Y.X. Saturated output feedback control for global asymptotic attitude tracking of spacecraft. *J. Guid. Control Dyn.* **2018**, *41*, 2300–2307. [CrossRef]
10. Xia, Y.Q.; Su, Y.X. Global saturated velocity-free finite-time control for attitude tracking of spacecraft. *IET Control Theory Appl.* **2019**, *13*, 1591–1602. [CrossRef]
11. Lee, D.; Leeghim, H. Reaction wheel fault-tolerant finite-time control for spacecraft attitude tracking without unwinding. *Int. J. Robust Nonlinear Control* **2020**, *30*, 3672–3691. [CrossRef]
12. Wang, Z.; Ma, J.; Wang, C. Predefined-time guaranteed performance attitude tracking control for uncertain rigid spacecraft. *IET Control Theory Appl.* **2022**, *16*, 1807–1819. [CrossRef]
13. Wang, Z.; Hu, L.M.; Ma, J. Distributed predefined-time attitude consensus control for multiple uncertain spacecraft formation flying. *IEEE Access* **2022**, *10*, 108848–108858. [CrossRef]
14. Slotine, J.J.E.; Li, W. *Applied Nonlinear Control*, 1st ed.; Prentice Hall: Englewood Cliffs, NJ, USA, 1991; pp. 41–126.
15. Sanfelice, R.G.; Goebel, R.; Teel, A.R. Invariance principles for hybrid systems with connections to detectability and asymptotic stability. *IEEE Trans. Autom. Control* **2007**, *52*, 2282–2297. [CrossRef]

Disclaimer/Publisher's Note: The statements, opinions and data contained in all publications are solely those of the individual author(s) and contributor(s) and not of MDPI and/or the editor(s). MDPI and/or the editor(s) disclaim responsibility for any injury to people or property resulting from any ideas, methods, instructions or products referred to in the content.

Article

Cooperative Guidance Strategy for Active Spacecraft Protection from a Homing Interceptor via Deep Reinforcement Learning

Weilin Ni [1], Jiaqi Liu [2], Zhi Li [1], Peng Liu [2] and Haizhao Liang [1,*]

[1] School of Aeronautics and Astronautics, Sun Yat-sen University, Shenzheng 518107, China; niwlin@mail2.sysu.edu.cn (W.N.); lizh336@mail2.sysu.edu.cn (Z.L.)
[2] National Key Laboratory of Science and Technology on Test Physics and Numerical Mathematics, Beijing 100076, China; liujiaqi_business@163.com (J.L.); lppl2008@163.com (P.L.)
* Correspondence: lianghch5@mail.sysu.edu.cn; Tel.: +86-133-211-967-99

Abstract: The cooperative active defense guidance problem for a spacecraft with active defense is investigated in this paper. An engagement between a spacecraft, an active defense vehicle, and an interceptor is considered, where the target spacecraft with active defense will attempt to evade the interceptor. Prior knowledge uncertainty and observation noise are taken into account simultaneously, which are vital for traditional guidance strategies such as the differential-game-based guidance method. In this set, we propose an intelligent cooperative active defense (ICAAI) guidance strategy based on deep reinforcement learning. ICAAI effectively coordinates defender and target maneuvers to achieve successful evasion with less prior knowledge and observational noise. Furthermore, we introduce an efficient and stable convergence (ESC) training approach employing reward shaping and curriculum learning to tackle the sparse reward problem in ICAAI training. Numerical experiments are included to demonstrate ICAAI's real-time performance, convergence, adaptiveness, and robustness through the learning process and Monte Carlo simulations. The learning process showcases improved convergence efficiency with ESC, while simulation results illustrate ICAAI's enhanced robustness and adaptiveness compared to optimal guidance laws.

Keywords: cooperative guidance; reinforcement learning; active protection; guidance law

MSC: 93-08

Citation: Ni, W.; Liu, J.; Li, Z.; Liu, P.; Liang, H. Cooperative Guidance Strategy for Active Spacecraft Protection from a Homing Interceptor via Deep Reinforcement Learning. *Mathematics* 2023, 11, 4211. https://doi.org/10.3390/math11194211

Academic Editor: Jiangping Hu

Received: 29 August 2023
Revised: 27 September 2023
Accepted: 7 October 2023
Published: 9 October 2023

Copyright: © 2023 by the authors. Licensee MDPI, Basel, Switzerland. This article is an open access article distributed under the terms and conditions of the Creative Commons Attribution (CC BY) license (https://creativecommons.org/licenses/by/4.0/).

1. Introduction

Spacecraft such as satellites, space stations, and space shuttles play an important role in both civil and military activities. They are also at risk of being intercepted in the exo-atmosphere. The pursuit-evasion game between the spacecraft and the interceptor will be critical in the competition for space resources and has been widely studied in recent years. The trajectory of spacecraft can be accurately predicted [1] since the dynamics of the spacecraft is generally described in terms of a two-body problem. With the development of accurate sensors, guidance technology, small-sized propulsion systems, and fast servo-mechanism techniques, the Kinetic Kill Vehicle (KKV), which can be used for direct-hit killing, has superior maneuverability compared to the other spacecraft. In other words, it is not practical for targeted spacecraft involved in the pursuit-evasion game to rely solely on orbital maneuvering.

Among the many available countermeasures, launching an Active Defense Vehicle (ADV) as a defender to intercept the incoming threat has proven to be an effective approach to compensate for the inferior target maneuverability [2–4]. In an initial study [2], Boyell proposed the active defense strategy of launching a defensive missile to protect the target from a homing missile. Boyell proposed an approximate normalized curve of game results under the condition of constant or static target velocity based on the relative motion relationship among the three participants. The dynamic three-body framework was introduced by Rusnak in Ref. [4], inspired by the narrative of a "lady-bodyguard-bandit" situation. This framework was later transformed into a "target-interceptor-defender" (TID) three-body spacecraft active defense game scenario as described in Ref [3]. In the TID scenario, the defender aims to reduce the distance from the interceptor, while the interceptor endeavors to increase the distance from the defender and successfully intercept the target. In Refs. [3,4], Rusnak proposed a game guidance method under the TID scenario based on Multiple Objective Optimization and differential games theories. It was proven that the proposed active defense method significantly reduces the miss distance and the required acceleration level between interceptor and defender.

The efficacy of the active defense method has garnered increased attention to the collaborative strategy between the target and defender in the TID scenario. Traditional methods for solving optimal strategies in this context include Optimal Control [5–7] and differential games theories [8–10]. In Ref. [7], Weiss employed the Optimal Control theory to independently design the guidance for both the target and defender. This approach considered the influence of target maneuvers on the interceptor's effectiveness as a defender. Furthermore, in Ref. [6], collaborative game strategies for the target and defender were proposed, emphasizing their combined efforts in the TID scenario. Aiming at the multi-member TID scenario in which a single target carries two defenders against two interceptors, Ref. [5] designed a multi-member cooperative game guidance strategy and considered the fuel consumption of target and defender. However, Optimal-Control-based strategies rely on perfect information, demanding accurate maneuvering details of the interceptor. In contrast, Differential Game approaches require prior knowledge instead of accurate target acceleration information, enhancing algorithm robustness [11]. In Ref. [8], optimal cooperative pursuit and evasion strategies were proposed using Pontryagin's minimum principle. A similar scenario was studied in Ref. [9] for both continuous and discrete domains using the linear–quadratic differential game method. It is worth noting that the differential game control strategies proposed in Ref. [9] solve the fuel cost and saturation problem. However, they introduce computational problems and make the selection of weight parameters more difficult. A switching surface [10], designed with zero-effort miss distance, was introduced to divide the multi-agent engagement into two one-on-one differential games, thereby achieving a balance between performance and usability. Nonetheless, using the differential game method to solve the multi-agent pursuit-evasion game problem still faces shortcomings [11–13]. First, it is difficult to establish a scene model of a multi-member, multi-role game due to the extremely large increase in the dimension of the state quantity; second, it has high requirements for the accuracy of the prior knowledge, and the success rate of the game is low if the prior knowledge of the players in the game cannot be obtained accurately; third, the differential game algorithm is complicated, involving a high-dimensional matrix operation, power function operation, integral calculation, etc., which places a high demand on the computational resources of the spacecraft. More on this topic can be found in [14–20].

With the advancement of machine learning technology, Deep Reinforcement Learning (DRL) has emerged as a promising approach for addressing active defense guidance problems. In DRL, an agent interacts with the environment and receives feedback in the form of rewards, enabling it to improve its performance and achieve specific tasks. This mechanism has led to successful applications of DRL in various decision-making domains, including robot control, MOBA games, autonomous driving, and navigation [21–25]. In Ref. [26], the DRL was utilized to learn an adaptive homing phase control law, accounting for sensor and actuator noise and delays. Another work [27] proposed an adaptive guidance system to address the landing problem using Reinforcement Meta-Learning, adapting agent training from one environment to another with limited steps, showcasing robust policy optimization in the presence of parameter uncertainties. In the context of the TID scenario, Lau [28] demonstrated the potential of using reinforcement learning for active defense guidance rating, although an optimal strategy was not obtained in their preliminary investigation.

It is worthy to point out that, on one hand, to better align with real-world engineering applications, research in guidance methods often needs to consider the presence of various information gaps and noise [29,30]. However, most of the existing optimal active defense guidance methods rely on perfect information assumptions, leading to subpar performance when faced with unknown prior knowledge or observation noise. Additionally, these methods often struggle to meet the real-time requirements of spacecraft applications. On the other hand, the majority of reinforcement learning algorithms have been applied to non-adversarial or weak adversarial flight missions, where mission objectives and process rewards are clear and intuitive. However, in the highly competitive TID game scenario, obtaining effective reward information becomes challenging due to the intense confrontation between agents, leading to sparse reward problems or "Plateau Phenomenon" [31].

Given these observations, there is a strong motivation to develop an active defense guidance method based on reinforcement learning that possesses enhanced real-time capabilities, adaptiveness, and robustness, while addressing the challenges posed by adversarial scenarios and sparse reward issues.

In this paper, we focus on the cooperative active defense guidance strategy design of a target spacecraft with active defense attempting to evade an interceptor in space. This TID scenario holds significant importance in the domains of space attack-defense and ballistic missile penetration. The paper begins by deriving the kinematic and first-order dynamic models of the engagement scenario. Subsequently, an intelligent cooperative active defense (ICAAI) guidance method for active defense is proposed, utilizing the twin-delay deep deterministic policy gradient (TD3) algorithm. To address the challenge of sparse rewards, an efficient and stable convergence (ESC) training approach is introduced. Furthermore, benchmark comparisons are made using Optimal Guidance Laws (OGLs), and simulation analyses are presented to validate the performance of the proposed method.

The paper is organized as follows. In Section 2, the problem formulation is provided. In Section 3, the guidance law is developed. In Section 4, experiments are presented where the proposed method has been compared with its analytical counterpart, followed by the conclusions presented in Section 5.

2. Problem Formulation

Consider a multi-agent game with a spacecraft as the main target (T), an active defense vehicle as the defender (D), and a highly maneuverable small spacecraft as the interceptor (I). In this battle, the interceptor chases the target, which launches the defender to protect itself by destroying the interceptor. During the endgame, all players are considered as constant-speed mass points whose trajectories can be linearized around the initial line of sight. As a consequence of trajectory linearization, the engagement, a three-dimensional process, can be simplified and will be analyzed in one plane. However, it should be noted that in most cases these assumptions do not affect the generality of the results [11].

A schematic view of the engagement is shown in Figure 1, where $X - O - Y$ is a Cartesian inertial reference frame. The distances between the players are denoted as ρ_{ID} and ρ_{IT}, respectively. Each player's velocity is indicated as V_I, V_T, and V_D, while their accelerations are represented as a_I, a_T, and a_D. The flight path angles of the players are defined as ϕ_I, ϕ_T, and ϕ_D, respectively. The line of sight (LOS) between the players is described by LOS_{ID} and LOS_{IT}, and the angles between the LOS and the X-axis are denoted as λ_{ID} and λ_{IT}. The lateral displacements of each player relative to the X-axis are represented as y_I, y_T, and y_D, while the relative displacements between the players are defined as y_{IT} and y_{ID}.

Considering the collective mission objectives, the target's priority is to evade the interceptor with defender support. Simultaneously, the interceptor aims to avoid the defender while chasing the target. Consequently, the target's guidance law strives for maximum convergence, while the defender's aims for convergence to zero. Conversely, the interceptor's guidance law assumes the opposite role (as depicted in Figure 1). This scenario can thus be segmented into two collision triangles: one involving the interceptor and the target, and the other between the interceptor and the defender.

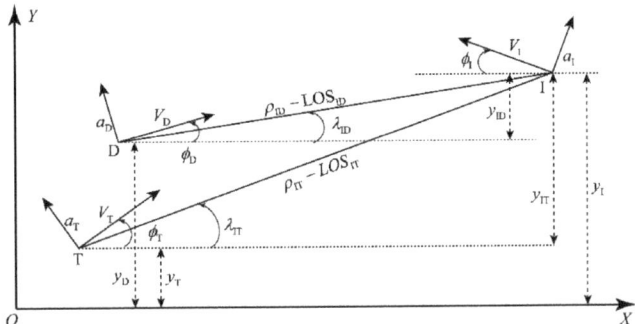

Figure 1. Schematic view of the engagement.

2.1. Equations of Motion

Consider the I-T collision triangle and the I-D collision triangle in a multi-agent pursuit-evasion engagement. The kinematics are expressed using the polar coordinate system attached in the target and defender as follows:

$$
\begin{aligned}
\dot{\rho}_{IT} &= -V_I \cos(\phi_I + \lambda_{IT}) - V_T \cos(\phi_T - \lambda_{IT}) \\
\dot{y}_{IT} &= V_I \sin \phi_I - V_T \sin \phi_T \\
\dot{\lambda}_{IT} &= \frac{V_I \sin(\phi_I + \lambda_{IT}) - V_T \sin(\phi_T - \lambda_{IT})}{\rho_{IT}}
\end{aligned} \quad (1)
$$

$$
\begin{aligned}
\dot{\rho}_{ID} &= -V_I \cos(\phi_I + \lambda_{ID}) - V_D \cos(\phi_D - \lambda_{ID}) \\
\dot{y}_{ID} &= V_I \sin \phi_I - V_D \sin \phi_D \\
\dot{\lambda}_{ID} &= \frac{V_I \sin(\phi_I + \lambda_{ID}) - V_D \sin(\phi_D - \lambda_{ID})}{\rho_{ID}}
\end{aligned} \quad (2)
$$

Furthermore, the flight path angles associated with dynamics can be defined for each of the players:

$$\dot{\phi}_i = \frac{a_i}{V_i}, \ i = \{I, T, D\} \quad (3)$$

2.2. Linearized Equations of Motion

In the research context, both the LOS angle λ and fight path angle ϕ are small quantities, and the inter-spacecraft distances are much larger than the spacecraft velocities. Furthermore, during the terminal guidance phase, the rate of change in spacecraft velocity magnitude approaches zero. Therefore, the equations of motion can be linearized around the initial line-of-sight:

$$
\begin{aligned}
\dot{\rho}_{IT} &= -V_I \cos(\phi_I + \lambda_{IT}) - V_T \cos(\phi_T - \lambda_{IT}) \approx -(V_I + V_T) \\
\ddot{y}_{IT} &= (V_I \sin \phi_I - V_T \sin \phi_T)' \approx (V_I \phi_I - V_T \phi_T)' \\
&= \dot{V}_I \phi_I - \dot{V}_T \phi_T + V_I \dot{\phi}_I - V_T \dot{\phi}_T = \dot{V}_I \phi_I - \dot{V}_T \phi_T + a_I - a_T \\
&\approx a_I - a_T \\
\dot{\lambda}_{IT} &= \frac{V_I \sin(\phi_I + \lambda_{IT}) - V_T \sin(\phi_T - \lambda_{IT})}{\rho_{IT}} \approx \frac{V_I(\phi_I + \lambda_{IT}) - V_T(\phi_T - \lambda_{IT})}{\rho_{IT}} \approx 0
\end{aligned} \quad (4)
$$

$$
\begin{aligned}
\dot{\rho}_{ID} &= -V_I \cos(\phi_I + \lambda_{ID}) - V_D \cos(\phi_D - \lambda_{ID}) \approx -(V_I + V_D) \\
\ddot{y}_{ID} &= (V_I \sin \phi_I - V_D \sin \phi_D)' \approx (V_I \phi_I - V_D \phi_D)' \\
&= \dot{V}_I \phi_I - \dot{V}_D \phi_D + V_I \dot{\phi}_I - V_D \dot{\phi}_D = \dot{V}_I \phi_I - \dot{V}_D \phi_D + a_I - a_D \\
&\approx a_I - a_D \\
\dot{\lambda}_{ID} &= \frac{V_I \sin(\phi_I + \lambda_{ID}) - V_D \sin(\phi_D - \lambda_{ID})}{\rho_{ID}} \approx \frac{V_I(\phi_I + \lambda_{ID}) - V_D(\phi_D - \lambda_{ID})}{\rho_{ID}} \approx 0
\end{aligned} \quad (5)
$$

The dynamics for each of the players is assumed to be a first-order process:

$$\dot{a}_i = -\frac{a_i - u_i}{\tau_i}, \quad i = \{I, T, D\} \tag{6}$$

Furthermore, the variable vector can be defined as follows:

$$x = \begin{bmatrix} y_{IT} & \dot{y}_{IT} & y_{ID} & \dot{y}_{ID} & a_I & a_T & a_D \end{bmatrix} \tag{7}$$

while the linearized equations of motion in the state space form can be written as follows:

$$\dot{x} = Ax + B \begin{bmatrix} u_I & u_T & u_D \end{bmatrix}^T \tag{8}$$

where

$$A = \begin{bmatrix} 0 & 1 & 0 & 0 & 0 & 0 & 0 \\ 0 & 0 & 0 & 0 & 1 & -1 & 0 \\ 0 & 0 & 0 & 1 & 0 & 0 & 0 \\ 0 & 0 & 0 & 0 & 1 & 0 & -1 \\ 0 & 0 & 0 & 0 & -1/\tau_I & 0 & 0 \\ 0 & 0 & 0 & 0 & 0 & -1/\tau_T & 0 \\ 0 & 0 & 0 & 0 & 0 & 0 & -1/\tau_D \end{bmatrix} \tag{9}$$

$$B = \begin{bmatrix} 0_{5 \times 3} \\ B_1 \end{bmatrix}, \quad B_1 = \begin{bmatrix} 1/\tau_I & 0 & 0 \\ 0 & 1/\tau_T & 0 \\ 0 & 0 & 1/\tau_D \end{bmatrix} \tag{10}$$

Since the velocity of each player is assumed to be constant, the engagement can be formulated as a fixed-time process. Thus, the interception time can be calculated using the following:

$$\begin{aligned} t_{f,IT} &= -\rho_{IT}^0/\dot{\rho}_{IT} = \rho_{IT}^0/(V_I + V_T) \\ t_{f,ID} &= -\rho_{ID}^0/\dot{\rho}_{ID} = \rho_{ID}^0/(V_I + V_D) \end{aligned} \tag{11}$$

where ρ_{IT}^0 represents the initial relative distance between the interceptor and the target, while ρ_{ID}^0 is the distance between the interceptor and the defender, allowing us to define the time-to-go of each engagement by

$$\begin{aligned} t_{go,IT} &= t_{f,IT} - t \\ t_{go,ID} &= t_{f,ID} - t \end{aligned} \tag{12}$$

which represents the expected remaining game time for the interceptor in the "Interceptor vs. Target" and "Interceptor vs. Defender" game scenarios, respectively.

2.3. Zero-Effort Miss

A well-known zero-effort miss (ZEM) is introduced in the guidance law design and reward function design. It is obtained from the homogeneous solutions of equations of motion and is only affected by the current state and interception time. It can be calculated as follows:

$$\begin{aligned} Z_{IT}(t) &= L_1 \Phi(t, t_{f,IT}) x(t) \\ Z_{ID}(t) &= L_2 \Phi(t, t_{f,ID}) x(t) \end{aligned} \tag{13}$$

where

$$\begin{aligned} L_1 &= \begin{bmatrix} 1 & 0 & 0 & 0 & 0 & 0 & 0 \end{bmatrix} \\ L_2 &= \begin{bmatrix} 0 & 0 & 1 & 0 & 0 & 0 & 0 \end{bmatrix} \end{aligned} \tag{14}$$

Thus, the ZEM and its derivative with respect to time are given as follows:

$$\begin{aligned} Z_{IT}(t) &= x_1 + t_{goIT} x_2 + a_I \tau_I^2 \varphi(t_{goIT}/\tau_I) x_5 - a_T \tau_T^2 \varphi(t_{goIT}/\tau_T) x_6 \\ Z_{ID}(t) &= x_3 + t_{goID} x_4 + a_I \tau_I^2 \varphi(t_{goID}/\tau_I) x_5 - a_D \tau_D^2 \varphi(t_{goID}/\tau_D) x_7 \end{aligned} \tag{15}$$

$$\dot{Z}_{\text{IT}}(t) = \tau_I \varphi(t_{\text{goIT}}/\tau_I) u_I - \tau_T \varphi(t_{\text{goIT}}/\tau_T) u_T$$
$$\dot{Z}_{\text{ID}}(t) = \tau_I \varphi(t_{\text{goID}}/\tau_I) u_I - \tau_D \varphi(t_{\text{goID}}/\tau_D) u_D \tag{16}$$

where

$$\varphi(\chi) = e^{-\chi} + \chi - 1 \tag{17}$$

2.4. Problem Statement

This research focuses on the terminal guidance task of evading a homing interceptor for a maneuvering target with active defense. We design a cooperative active defense guidance to facilitate coordinated maneuvers between the target and the defender based on DRL. This enables the target to evade the interceptor's interception while allowing the defender to counter-intercept the incoming threat.

3. Guidance Law Development

In this section, we develop the Intelligent Cooperative Active Defense (ICAAI) guidance strategy and design an efficient and stable convergence (ESC) training approach. The target and defender utilize ICAAI guidance, while the interceptor employs OGL. We describe the game scenario using a Markov process, present the ICAAI guidance strategy, and design an ESC training approach based on reward shaping and curriculum learning.

3.1. Markov Decision Process

The sequential decision making that an autonomous RL agent interacts with the environment (e.g., the engagement) can be formally described as an MDP, which is required to properly set up the mathematical framework of an DRL problem. A generic time-discrete MDP can be represented as a 6-tuple $\{s, o, a, P_{sa}, \gamma, R\}$. $s_t \in S \in \mathbb{R}^n$ is a vector that completely identifies the state of the system (e.g., the EOM) at time t. Generally, the complete state is not available to the agent at each time t; the decision-making relies on an observation vector $o_t \in O \in \mathbb{R}^m$. In the present paper, the observations are defined as an uncertain (e.g., imperfect and noisy) version of the true state, which can be written as a function Ω of the current state s_t. The action $a \in A \in \mathbb{R}^l$ of the agent is given by a state-feedback policy $\pi : O \to A$, that is, $a_t = \pi(o_t)$. P_{sa} is time-discrete dynamic model describing the transformation led by the state–action pair (s_t, a_t). As a result, the evolution rule of the dynamic system can be described as follows:

$$\begin{aligned} s_{t+1} &= P_{sa}(s_t, a_t) \\ o_t &= \Omega(s_t) \\ a_t &= \pi(o_t) \end{aligned} \tag{18}$$

Since a fixed-time engagement is considered, the interaction between the agent and the environment gives rise to a trajectory **I**:

$$\begin{aligned} \mathbf{I} &= [\iota_1, \iota_2, \cdots, \iota_t, \cdots, \iota_{T-1}, \iota_T] \\ \iota_t &= [o_t, a_t, r_t]^T \end{aligned} \tag{19}$$

where the trajectory information at each time step ι_t is composed of observational o_t, action a_t, and reward signal r_t generated through the interaction between the agent and the environment.

The return, the agent received at time t in the trajectory **I**, is defined as a discounted sum of rewards:

$$R_t^{\mathbf{I}} = \sum_{i=t}^{T} \gamma^{i-t} r_i \tag{20}$$

where $\gamma \in (0, 1]$ is a discount rate determining whether the agent has a long-term vision ($\gamma = 1$) or is short-sighted ($\gamma \ll 1$).

Prior to deriving the current guidance law, we outline the key elements of the MDP: state space, action space, and observations. We present the reward design separately by highlighting a crucial aspect of the configuration.

3.1.1. Perfect Information Model

In a deterministic model, the basic assumption is that each player has perfect information about the interceptor (e.g., states, maximum acceleration, and time constant). The communication of this information between the defender and the protected target is assumed to be ideal and without delay. Thus, the state space can be identified by states, maximum acceleration, and time constant:

$$s_t = [t \quad x_t \quad y_t \quad V_t \quad a_t \quad a_{max} \quad \tau]^T \quad (21)$$

$$x_t = [x_{t,T} \quad x_{t,D} \quad x_{t,I}], y_t = [y_{t,T} \quad y_{t,D} \quad y_{t,I}] \quad (22)$$

$$V_t = [V_{t,T} \quad V_{t,D} \quad V_{t,I}] \quad (23)$$

$$a_t = [a_{t,T} \quad a_{t,D} \quad a_{t,I}] \quad (24)$$

$$a_{max} = [a_{max,T} \quad a_{max,D} \quad a_{max,I}] \quad (25)$$

$$\tau = [\tau_T \quad \tau_D \quad \tau_I] \quad (26)$$

As with the multi-agent system, interactions introduce uncertainty into the environment, which significantly affects the stability of the RL algorithm. Given the full cooperation between defender and target due to communication assumptions, the model must learn a shared guidance law for both. This effectively mitigates environmental uncertainty and enhances model convergence. In practical application, the same trained agent is assigned to the target pair, yielding the following action space:

$$\text{action} = [u_T \quad u_D] \quad (27)$$

Since the dynamics of the scenario are formulated in Section 2.1, the state can be propagated implicitly as the linearized equation of motion presented in Equations (4)–(6).

3.1.2. Imperfect Information Model

The imperfection of information is usually due to the limitations of radar measurement and the erasure of prior knowledge. However, in existing studies, perfect information is a strong assumption, which leads to implementation difficulties in practice. To address this dilemma, this thesis considers information degradation. On the one hand, the interceptor is assumed to have perfect information (i.e., the relative states and maneuverability of the target and the defender). On the other hand, the observation of the target and defender is imperfect and even noise-corrupted. The observation uncertainty is modeled as observation noise and a mask on the perfect information.

$$o_t = \Omega(s_t) = \Gamma s_t \times (I + \omega_{o,t}) = \begin{bmatrix} t \\ x_t \\ y_t \\ V_t \\ a_t \end{bmatrix} + \begin{bmatrix} 0 \\ \delta x_{o,t} \\ \delta y_{o,t} \\ \delta V_{o,t} \\ \delta a_{o,t} \end{bmatrix} \quad (28)$$

where Γ is the mask matrix and $\omega_{o,t}$ is the observation noise vector. $\omega_{o,t}$ can be calculated by Equations (29)–(32).

$$\omega_{o,t} = \begin{bmatrix} 0 \\ \delta x_{o,t} \\ \delta y_{o,t} \\ \delta V_{o,t} \\ \delta a_{o,t} \end{bmatrix} \sim \mathcal{U}(0_{13}, \Sigma) \in \mathbb{R}^{13} \tag{29}$$

$$\Sigma = [0, 0, 0, \sigma_{xI}, 0, 0, \sigma_{yI}, 0, 0, \sigma_v, 0, 0, \sigma_a]^T \tag{30}$$

$$\sigma_{xI} = \cos(\sigma_{LOS} + \lambda_{IT})(\rho_{IT} + \sigma_\rho) - x_I \approx 0 \tag{31}$$

$$\sigma_{yI} = \sin(\sigma_{LOS} + \lambda_{IT})(\rho_{IT} + \sigma_\rho) - y_I \approx \sigma_{LOS} \cdot \rho_{IT} \tag{32}$$

where Σ represents the noise amplitude, with $\sigma_\rho(m)$, $\sigma_{LOS}(mrad)$, $\sigma_v(m/s)$, and $\sigma_a(m/s^2)$ the nonnegative parameters.

3.2. ICAAI Guidance Law Design

In this section, we present the mathematical framework of actor–critic RL algorithms, focusing on the algorithm used in ICAAI guidance: Twin-Delay Deep Deterministic Policy Gradient (TD3) [32]. TD3 is an advanced deterministic policy gradient reinforcement learning algorithm. In comparison to stochastic policy gradient algorithms like Proximal Policy Optimization (PPO) [33] and Asynchronous Advantage Actor–Critic (A3C) [34], TD3 exhibits a higher resistance to converging into local optima. Furthermore, when compared to traditional deterministic policy gradient RL algorithms such as Deep Deterministic Policy Gradient (DDPG) [35], TD3 achieves superior training stability and convergence efficiency. This assertion is supported by our prior RL algorithm selection experiments, as illustrated in Figure 2.

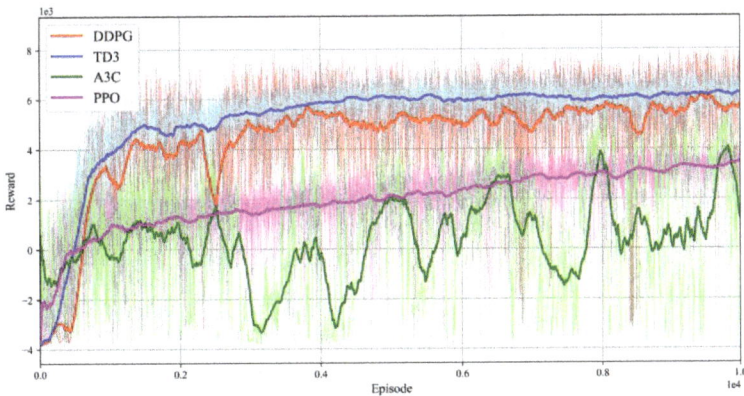

Figure 2. Comparison of training results of various reinforcement learning algorithms.

Without loss of generality, throughout the entire section the MDP is supposed to be perfectly observable (i.e., with $o_t = s_t$) to conform with the standard notation of RL. However, the perfect information state s_t can be replaced by observation o_t whenever the observations differ from the state.

3.2.1. Actor–Critic Algorithms

The RL problem's goal is to find the optimal policy π_ϕ with parameters ϕ that maximizes the expected return, which can be formulated as follows:

$$J(\phi) = \mathop{\mathbb{E}}_{\tau \sim \pi_\phi}[R_0^\tau] = \mathop{\mathbb{E}}_{\tau \sim \pi_\phi}\left[\sum_{i=0}^{T} \gamma^{i-0} r_i\right] \tag{33}$$

where $\mathop{\mathbb{E}}_{\tau \sim \pi}$ denotes the expectation taken over the trajectory τ. In actor–critic algorithms the policy, known as the actor, can be updated by using a deterministic policy gradient algorithm [36]:

$$\nabla_\phi J(\phi) = \mathbb{E}_{P_{sa}}\left[\nabla_a Q^\pi(s,a)\big|_{a=\pi(s)} \nabla_\phi \pi_\phi(s)\right] \tag{34}$$

The expected return, when performing action a in state s and following π after, is called the critic or the value function, which can be formulated as follows:

$$Q^\pi(s,a) = \mathop{\mathbb{E}}_{\tau \sim \pi_\phi}[R_t^\tau|s,a] \tag{35}$$

The value function can be learned through off-policy temporal differential learning, an update rule based on the Bellman equation which describes the relationship between the value of the state–action pair (s,a) and the value of the subsequent state–action pair (s',a'):

$$Q^\pi(s,a) = r + \gamma \mathop{\mathbb{E}}_{s',a'}[Q^\pi(s',a')] \tag{36}$$

In deep Q-learning [37], the value function can be estimated with a neural network approximator $Q_\theta(s,a)$ with parameters θ, and the network is updated by using temporary differential learning with a secondary frozen target network $Q_{\theta'}(s,a)$ to maintain a fixed objective U over multiple updates:

$$U = r + \gamma Q_{\theta'}(s',a'), \ a' = \pi_{\phi'}(s') \tag{37}$$

where the actions a' are determined by a target actor network $\pi_{\phi'}$. Generally, the loss function and update rule can be formulated as follows:

$$J(\theta) = U - Q_\theta(s,a) \tag{38}$$

$$\nabla_\theta J(\theta) = [U - Q_\theta(s,a)] \nabla_\theta Q_\theta(s,a) \tag{39}$$

The parameters of target networks are updated periodically to exactly match the parameters of the corresponding current networks, which is called delayed update. This leads to the original actor–critic method, the basic structure of which is shown in Figure 3.

3.2.2. Twin-Delayed Deep Deterministic Policy Gradient Algorithm

To address the common RL issues in actor-critic algorithms (i.e., overestimation bias and accumulation of errors), in the TD3 algorithm, the actor–critic framework is modified from three aspects.

A novel variant of double Q-learning [38] called clipped double Q-learning is developed to limit possible overestimation. This provides the update objective of the critic:

$$U = r + \gamma \min_{i=1,2} Q_{\theta'_i}(s', \pi_{\phi'_1}(s')) \tag{40}$$

The parameters of policy networks are updated periodically to match the value network, which is called delayed policy update, and the soft update approach is adopted, which can be formulated as follows:

$$\theta' \leftarrow \kappa\theta + (1-\kappa)\theta' \tag{41}$$

where κ is a proportion parameter.

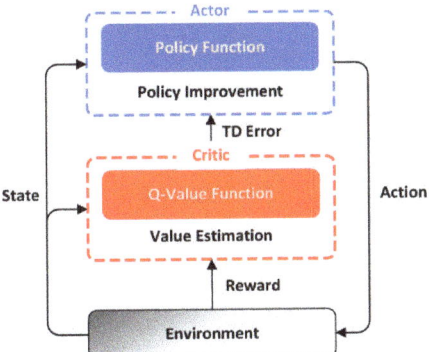

Figure 3. Structure of actor–critic method.

Target policy smoothing regularization is adopted to alleviate the overfitting phenomenon, which can be explicated as follows:

$$y = r + \gamma Q_{\theta'}(s', \pi_{\phi'}(s') + \varepsilon) \tag{42}$$

where ε is a clipped Gaussian noise.

An overview of the TD3 algorithm is demonstrated in Figure 4.

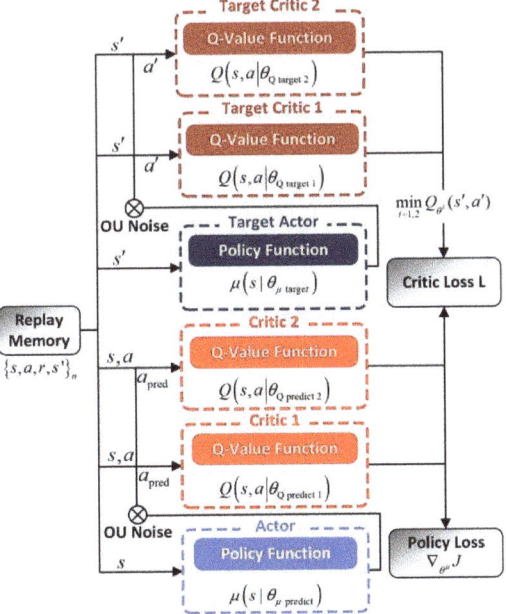

Figure 4. Structure of the TD3 algorithm.

3.2.3. Implementation Details

As for the network architecture setting, the agent observations are vectors with 13 dimensions. Both the guidance policy estimation (actor) and the value function estimation (critic) consist of three fully connected layers with sizes of 64, 256, and 512, respectively, along with layer normalization. The output layer has two units for the actor, representing the unified command of the target and defender, respectively, and one unit for the critic. The activation function is ReLU for the hidden layer neurons and linear for the output layer neuron. This structure is heuristically designed and can be generalized for efficient function approximation. Deeper and wider networks are avoided for real-time performance and fast convergence.

The hyperparameters of TD3 have been devised and validated by empirical experiments, which are reported in Table 1.

Table 1. TD3 hyperparameters.

Hyperparameter	Symbol	Value
Discount factor	γ	0.99
Learning rate	α	3×10^{-4}
Buffer size	$\mathbb{B}$	5120
Batch size	n_{batch}	128
Soft update coefficient	ζ	5×10^{-3}
Policy delay	n_{opt}	2
Train frequency	ω	6000

3.3. ESC Traning Technique

Aiming at the sparse reward problem in the multi-agent pursuit-evasion game, an efficient and stable convergence (ESC) training approach of reinforcement learning is proposed based on reward shaping [39] and curriculum learning [40].

3.3.1. Reward Shaping

The design of a reward function is the most challenging part of solving this multi-agent pursuit-evasion game through RL, as the function had to be adaptive to engagement with a sparse reward setting. It is found that, except for the common leadership mission, the pursuit-evasion game can be formulated as a strictly competitive zero-sum game. In addition, the agent policy network weights were randomly initiated at the beginning of training, while the interceptor was deployed with optimal guidance and is sufficiently aggressive.

In [41], a shaping technique was presented as a particularly effective approach to solving sparse reward problems through a series of biological experiments. The researchers divided a difficult task into several simple units and trained the animals according to an easy-to-hard schedule. This approach requires adjusting the reward signal to cover the entire training process, followed by gradual changes in task dynamics as training progresses. In [40], researchers took this idea further and proposed curriculum learning, a type of training strategy. In this work, the shaping technique and curriculum learning were used to speed up the convergence of neural networks and to increase the stability and performance of the algorithm.

The goal of the target and the defender is to converge Z_{ID} to zero as $t \to t_{f2}$ while keeping Z_{IT} as large as possible. On the contrary, the interceptor control law is designed to make Z_{IT} converge to zero while maintaining Z_{ID} as large as possible.

For this reason, a non-sparse reward function is defined in Equations (43) and (44):

$$r_{medium} = \gamma \Phi(s') - \Phi(s) \tag{43}$$

$$\Phi(s) = \left|\frac{Z_{\text{IT}}}{\alpha_1}\right|^{\beta_1} - \left|\frac{Z_{\text{ID}}}{\alpha_2}\right|^{\beta_2} \tag{44}$$

$$r_{\text{terminal}} = \begin{cases} \sigma, & \text{if succeed} \\ -\sigma, & \text{else} \end{cases} \tag{45}$$

where γ is the discount factor in the Markov decision process and α_1, α_2, β_1, β_2, and σ are the positive hyperparameters.

It must be stressed that, since both the number and maneuverability of players completely change the environment, the hyperparameter values used in this paper may be not universal. Thus, in the following subsection, the focus will be on the applied design method instead of the specific hyperparameter values. The r_{terminal} is the terminal reward signal given to the terminal behavior of the agent, which is sparse but intuitive. Situations in which the interceptor is destroyed by the defender (when $t = t_{f,\text{ID}}$) or when the interceptor is driven away by the defender and misses the target are judged as a success. Furthermore, the r_{medium} is a non-sparse reward function based on the difference form of a potential function $\Phi(s)$ which ensures the consistency of the optimal strategy [42–44]. It is important to emphasize that the design of $\Phi(s)$ relies on a fractional exponential function. This function provides a continuous reward signal for the agent's evaluation of each state. Notably, this model exhibits a unique property: as the base number approaches infinity, the gradient decreases to zero, and as the base number approaches zero, the gradient increases infinitely. This specific characteristic significantly aids the agent in converging towards states where the base number is either greater than zero or approaches zero.

In this paper, the defined reward function carries the physical meaning of the mission—the target must escape from the interceptor, while the defender has to get close to the interceptor. The r_{medium} value increases as Z_{ID} converges to zero, or when Z_{IT} increases. On the other hand, it decreases when Z_{ID} is divergent or when Z_{IT} converges to zero.

Generally, reward normalization is beneficial to neural network convergence. However, determining the bounds of Z_{IT} and Z_{ID} is a complex task. For this reason, hyperparameters α_1, α_2, β_1, and β_2 are tuned, aiming to scale the r_{medium} close to $[-c, c]$, in which c is a positive constant. In the following step, the design of ρ is considered, which introduces the expectation of agent foresight. If the agent is expected to predict the terminal reward r_{terminal} n steps before, the discounted terminal reward must be larger than the r_{medium} bounds. Thus, the hyperparameter ρ satisfies the following expression:

$$\rho \geq \frac{c}{\gamma^n} \tag{46}$$

3.3.2. Curriculum Learning

After hyperparameter tuning, we enhance the training stability of intelligent algorithms using an adaptive progressive curriculum learning approach. This method incrementally raises training complexity to enhance agent capability and performance. The agent's training level is adaptively assessed through changes in network loss, determining appropriate training difficulty. The v_{PG} calculation formula is as follows:

$$v_{\text{PG}} = L(x, \theta) - L(x, \theta') \tag{47}$$

where $L(\cdot)$ represents the calculation function of network loss; θ is the current network parameter and θ' is the new network parameter obtained after data x training. Given a small amount $\varepsilon(1 \gg \varepsilon > 0)$, when

$$|v_{\text{PG}}| < \varepsilon \tag{48}$$

the agent training enters the next stage. A sequence of increasingly difficult tasks is allocated to the agent, as shown in Table 2. The curriculum was divided into three stages:

- The agent is required to combat the interceptors employing non-maneuvering;

- Square wave signal;
- OGL.

Finally, it is possible to complete the reward shaping process.

Table 2. Curriculum learning.

Curriculum	Stage 1	Stage 2	Stage 3
Interceptor guidance command	None	Square wave signal	OGL
Maximum interceptor acceleration	0	8 g	4 g/6 g/8 g

In summary, the block diagram of ICAAI guidance strategy is shown in Figure 5.

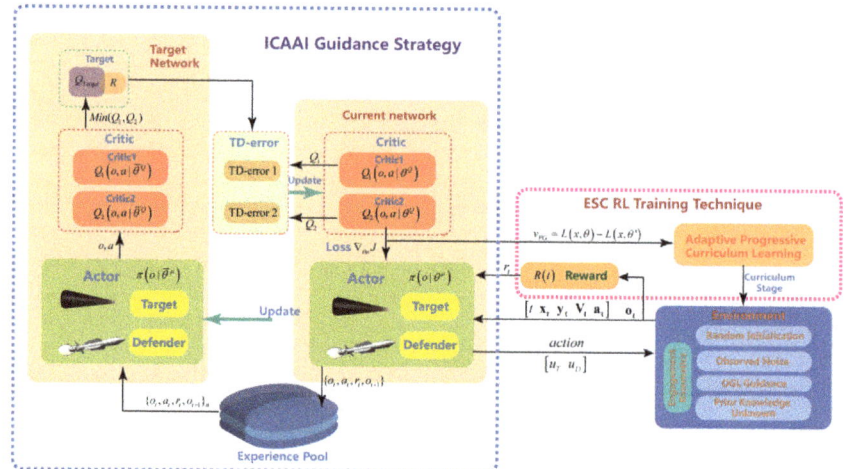

Figure 5. Block Diagram of ICAAI Guidance Strategy.

4. Experiments

In this section, we demonstrate the efficacy of the proposed guidance method and the effectiveness of the shaping technique through learning processes and Monte Carlo simulations. We establish benchmark comparisons by including OGLs and evaluating application requirements. To illustrate, we consider a scenario [10] involving a maneuverable small spacecraft (Interceptor, I), a defensive vehicle (Defender, D), and an evading spacecraft (Target, T), all in circular Earth orbits. Gravity effects are incorporated in the simulations. Assumptions include the interceptor's superior maneuverability and time constant compared to the target and defender.

4.1. Optimal Pursuit and Evasion Guidance Laws

Lemma 1. *The linear–quadratic optimal guidance law (LQOGL) [10]:*

$$u_I^* = \begin{cases} -\frac{K(t)Z_{\text{ID}}(t)}{\omega_1} u_I^{\max} \tau_I \varphi\left(\frac{t_{f,\text{ID}}-t}{\tau_I}\right) & \text{for } \|Z_{\text{ID}}(t)\| < \eta \\ -\frac{P(t)Z_{\text{IM}}(t)}{\xi_1} u_I^{\max} \tau_I \varphi\left(\frac{t_{f,\text{IT}}-t}{\tau_I}\right) & \text{else} \end{cases} \quad (49)$$

where η is a positive constant representing the limit-collision radius between the interceptor and the defender, and u_I^{max} is the maximum control force provided by the interceptor. Furthermore, variable $K(t)$ and $P(t)$ can be defined as follows:

$$K(t) = \frac{1}{\int_t^{t_{fID}} \left[\frac{1}{\omega_1}\left(u_I^{max}\tau_I\varphi\left(\frac{t_{f,ID}-t}{\tau_I}\right)\right)^2 - \frac{1}{\omega_2}\left(u_D^{max}\tau_D\varphi\left(\frac{t_{f,ID}-t}{\tau_D}\right)\right)^2\right]dt - 1} \quad (50)$$

$$P(t) = \frac{1}{\int_t^{t_{fIM}} \left[\frac{1}{\xi_1}\left(u_I^{max}\tau_I\varphi\left(\frac{t_{f,IM}-t}{\tau_I}\right)\right)^2 - \frac{1}{\xi_2}\left(u_M^{max}\tau_M\varphi\left(\frac{t_{f,IM}-t}{\tau_M}\right)\right)^2\right]dt - 1} \quad (51)$$

where ω_1, ω_2, ξ_1, and ξ_2 are nonnegative constants ensuring the interceptor converges towards the target, guaranteeing its escape from the defender.

Proof. The detailed proof of similar results can be found in [10]; see Theorem 1 and the associated proof. □

Lemma 2. *Standard optimal guidance law (SOGL) [45]:*

$$\begin{array}{l} u_I^* = u_I^{max}\text{sgn}[Z_{ID}(t_{f,ID})]\text{sgn}\left[\varphi\left(\frac{t_{f,ID}-t}{\tau_I}\right)\right] \text{ for } \|Z_{ID}(t)\| < \eta \\ u_I^* = -u_I^{max}\text{sgn}[Z_{IT}(t_{f,IT})]\text{sgn}\left[\varphi\left(\frac{t_{f,IT}-t}{\tau_I}\right)\right] \text{ else} \end{array} \quad (52)$$

where η is a positive constant representing the switching condition always equal to the defender kill radius.

Proof. Consider the following cost function:

$$\begin{array}{l} J_1 = -\frac{1}{2}Z_{ID}^2(t_{fID}) \text{ for } \|Z_{ID}(t)\| < \eta \\ J_2 = \frac{1}{2}Z_{IT}^2(t_{fIT}) \text{ else} \end{array} \quad (53)$$

For J_1, the Hamiltonian of the problem is defined as follows:

$$H_1 = \lambda_1 \dot{Z}_{ID}(t) \quad (54)$$

The costate equation and transversality condition are provided by the following:

$$\dot{\lambda}_1(t) = -\frac{\partial H_1}{\partial Z_{ID}} = 0 \quad (55)$$

$$\lambda_1(t_{fID}) = \frac{\partial J_1}{\partial Z_{ID}(t_{fID})} = -Z_{ID}(t_{fID}) \quad (56)$$

The optimal interceptor controller minimizes the Hamiltonian satisfying the following:

$$u_I^* = \underset{u_I}{\text{argmin}}(H_1) \quad (57)$$

The interceptor guidance law can thus be obtained:

$$u_I^* = u_I^{max}\text{sgn}[Z_{ID}(t_{fID})]\text{sgn}\left[\varphi\left(\frac{t_{fID}-t}{\tau_I}\right)\right] \quad (58)$$

For J_2, a similar interceptor guidance law can be found:

$$u_I^* = -u_I^{max}\text{sgn}\left[Z_{IT}\left(t_{fIT}\right)\right]\text{sgn}\left[\varphi\left(\frac{t_{fIT}-t}{\tau_I}\right)\right] \quad (59)$$

Finally, the interceptor guidance schemes for evading the defender and pursuing the target are proposed after combining Equations (58) and (59):

$$\begin{aligned} u_I^* &= u_I^{max} \text{sgn}\left[Z_{ID}\left(t_{fID}\right)\right] \text{sgn}\left[\varphi\left(\frac{t_{fID}-t}{\tau_1}\right)\right] \text{ for } \|Z_{ID}(t)\| < \eta \\ u_I^* &= -u_I^{max} \text{sgn}\left[Z_{IT}\left(t_{fIT}\right)\right] \text{sgn}\left[\varphi\left(\frac{t_{fIT}-t}{\tau_1}\right)\right] \text{ else} \end{aligned} \quad (60)$$

□

4.2. Engagement Setup

In this scenario, a target carrying an active anti-interceptor is threatened by a KKV interceptor in orbit at an altitude of 500 km. The defender maintains an initial safe distance of approximately 50 m longitudinally and 10 km transversely to the target. Given that the detection range of the interceptor's guided warhead is about 100 km, the initial transverse distance between the interceptor and the target is set at 100 km, and the initial longitudinal position is random in the range 499.8–500.2 km. In addition, the maneuverability and control response speed of the interceptor are better than those of the target and defender, and the OGL is used for guidance.

The comprehensive list of engagement parameters is shown in Table 3.

Table 3. Engagement parameters.

Parameters	Interceptor	Target	Defender
Horizontal location (km)	100	0	0~15
Vertical location (km)	499.8~500.2	500	500.05
Horizontal velocity (km/s)	−3	2	2
Vertical velocity	0	0	0
Maximum acceleration (g)	8	2	6
Time constant (s)	0.02	0.1	0.05
Kill radius (m)	0.25	0.5	0.15

Furthermore, Gaussian noise with standard variance of $\sigma_{LOS} = 1$ mrad, $\sigma_v = 0.2$ m/s, and $\sigma_a = 1$ m/s² is considered in the interceptor information obtained by the target and defender through a radar seeker.

4.3. Experiment 1: Real-Time Performance of the Guidance Policy

To verify that the proposed RL training approach ESC can improve convergence efficiency and stability, the learning processes were demonstrated using the sparse reward (SR) signal and ESC, respectively, with the same hyperparameters. During the learning process, the weights of the neural network model were stored every 100 episodes for subsequent analysis. In addition, to remove stochasticity as a confounding factor, six random seeds were set for each case. Meanwhile, the real-time performance of the optimized agent is evaluated by comparing it with the traditional OGLs.

The agents were obtained after a training of 20,000 episodes, which took 12 h with 8 parallel workers on a computer equipped with a 104-core Intel Core Xeon Platinum 8270 CPU @2.70 GHz. Similarly, both the traditional methods and the proposed method are provided a current state or observation and return the required action. Table 4 shows the comparison of computational cost and update frequency obtained by using SOGL, LQOGL, and the proposed method. It can be seen from the table that LQOGL is time-consuming due to the calculation of the Riccati function, which is the reason why it has not been applied in practice. As a proven approach, the SOGL has excellent real-time performance. The proposed method achieved an update frequency of 10^3 Hz and showed great potential for on-board applications. While a variety of approaches (e.g., pruning and distillation) were effective to compress the policy network and further improve its real-time performance, it is not the main work of this research.

Table 4. Statistics of time consumption with different guidance methods.

Metrics	LQOGL	SOGL	ICAAI
Duration (1e3 step)	2.773 s	0.0145 s	0.910 s
Update frequency	≈360 HZ	≈6.9×10^4 HZ	≈1.1×10^3 HZ

Remark 1. *As shown in Equations (18) and (19), the LQOGL has to solve the Riccati differential equation. However, the experimental results show that its update frequency cannot meet the real-time requirements of spacecraft guidance. Compared to the LQOGL, the SOGL in Equation (60) does not need to solve the Riccati differential equation and has no hyperparameter. This improves both its computational efficiency and robustness at the cost of flexibility and the occurrence of the chattering phenomenon. To take into account the practical situation, the SOGL was chosen as an OGL benchmark.*

4.4. Experiment 2: Convergence and Performance of the Guidance Policy

The performance of the trained agent in the fully observable game was investigated by comparing the escape success rate corresponding to an optimized policy $\pi_\phi(s)$, obtained by performing Monte Carlo simulation in the fully observable (deterministic and with default engagement parameters) environment, with the solution of the SOGL.

4.4.1. Baselines

The SOGL for the target and the defender were considered as an OGL benchmark. Through a brief derivation similar to that in Section 3, it can be proven that the SOGLs for the target and the defender are as follows:

$$u_T = -u_T^{max} \text{sgn}\left[Z_{IT}\left(t_{fIT}\right)\right] \text{sgn}\left[\varphi\left(\frac{t_{fIT}-t}{\tau_T}\right)\right]$$
$$u_D = u_D^{max} \text{sgn}\left[Z_{ID}\left(t_{fID}\right)\right] \text{sgn}\left[\varphi\left(\frac{t_{fID}-t}{\tau_D}\right)\right] \quad (61)$$

4.4.2. Convergence and Escape Success Rate

Figure 6 displays the learning curves depicting the mean accumulated reward across learning episodes for various scenarios. As depicted, in the ESC case, the agent's reward consistently escalated throughout the training episodes, ultimately stabilizing at around 6000 after 4000 iterations. Conversely, within the sparse reward (SR) framework, the ICAAI encountered a plateau phenomenon during training, resulting in an unstable convergence process for the associated reward function and eventual convergence failure.

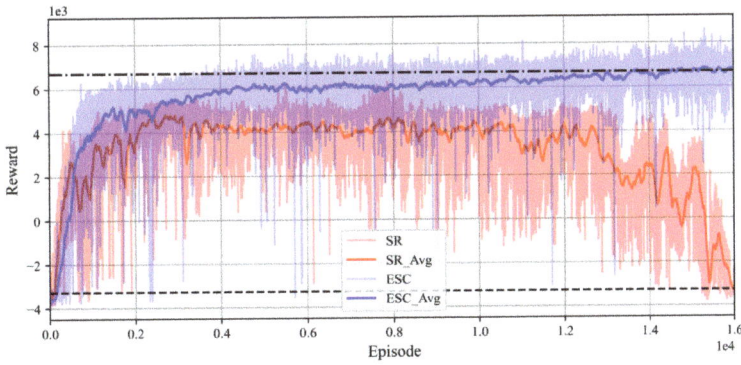

Figure 6. Learning curves of the ICAAI.

Figure 7 presents success rate curves for target evasion over learning episodes, comparing agents trained with and without ESC. The green line denotes OGL's deterministic environment success rate of 83.4%. The ESC-trained agent surpassed the baseline by 2700 episodes, achieving a peak performance of 99% after around 13,800 episodes. Conversely, the agent without ESC exhibited a gradual decline in performance after reaching a zenith of 77%, signifying policy network overfitting during continued training. The ESC-trained agent demonstrated accelerated convergence and improved local optima. It can be inferred that the proposed ESC training approach effectively organizes exploration, addressing sparse reward issues and showcasing heightened learning efficiency and asymptotic performance. Furthermore, the proposed methodology adeptly mitigates overfitting phenomena.

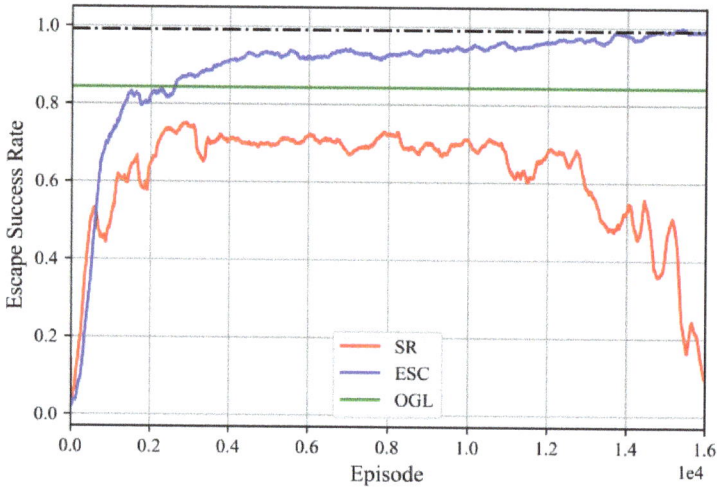

Figure 7. Escape success rate.

4.4.3. Performance Test

Figure 8 depicts spacecraft trajectories, featuring the interceptor's actual path (blue curve) and the observed trajectory from the target's perspective (yellow curve). Figure 9 displays the lateral acceleration profiles for each spacecraft, while Figure 10 illustrates the ZEM measurements between the target and interceptor and between the defender and interceptor. The simulation results presented in Figure 11 reaffirm the impact of the relative distance between the target and defender dis_{DT} on the game outcomes for the target.

Figures 8–10 illustrate the evident cooperation between the target and the defender, utilizing relative state information. Taking the simulation results at $dis_{DT} = 10$ km as an example, the miss distance between the target and the interceptor was approximately 15 m. The defender maintained a miss distance of less than 1 m from the interceptor, confirming its successful interception threat. Figures 9 and 10 depict that, within 16 s of the scenario's initiation, the target collaborated with the defender, executing subtle maneuvers to intercept the interceptor. At around the 16 s mark, the interceptor perceived the threat and initiated an escape strategy. Simultaneously, the target executed an evasive maneuver in the opposite direction, utilizing its maximum maneuverability, which resulted in an increase in distance. Ultimately, the interceptor managed to evade the defender's interception attempt but failed to intercept the target in time, leading to the target's successful evasion.

In addition, the above simulation results show that the relative distance between the target and defender dis_{DT} directly determines the time it takes for the interceptor to intercept the target after evading the defender. Consequently, dis_{DT} significantly influences the game outcomes for the target, including the success rate of evasion and miss distance. Therefore, to explore the effect of dis_{DT} on the performance of ICAAI, the game results for dis_{DT} ranging from 0 to 15 km are introduced in Figure 11.

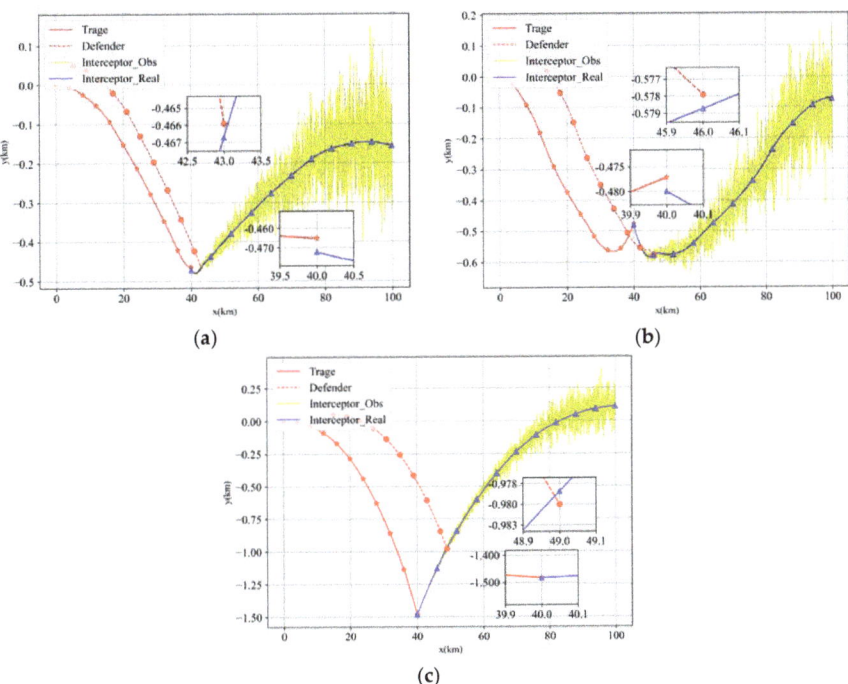

Figure 8. Spacecrafts game trajectory. (**a**) $dis_{DT} = 5$ km, (**b**) $dis_{DT} = 10$ km, (**c**) $dis_{DT} = 15$ km.

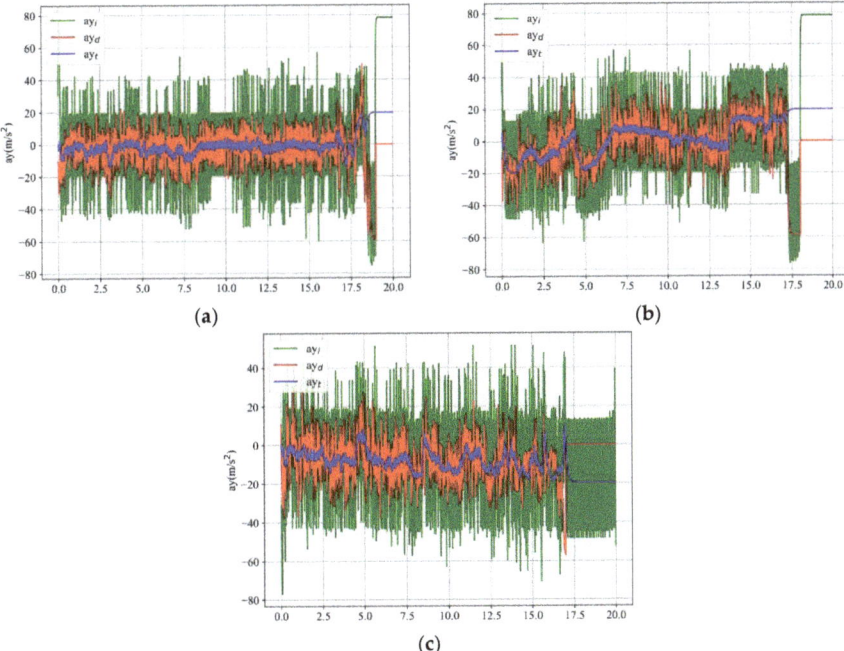

Figure 9. Lateral acceleration curve of each spacecraft. (**a**) $dis_{DT} = 5$ km, (**b**) $dis_{DT} = 10$ km, (**c**) $dis_{DT} = 15$ km.

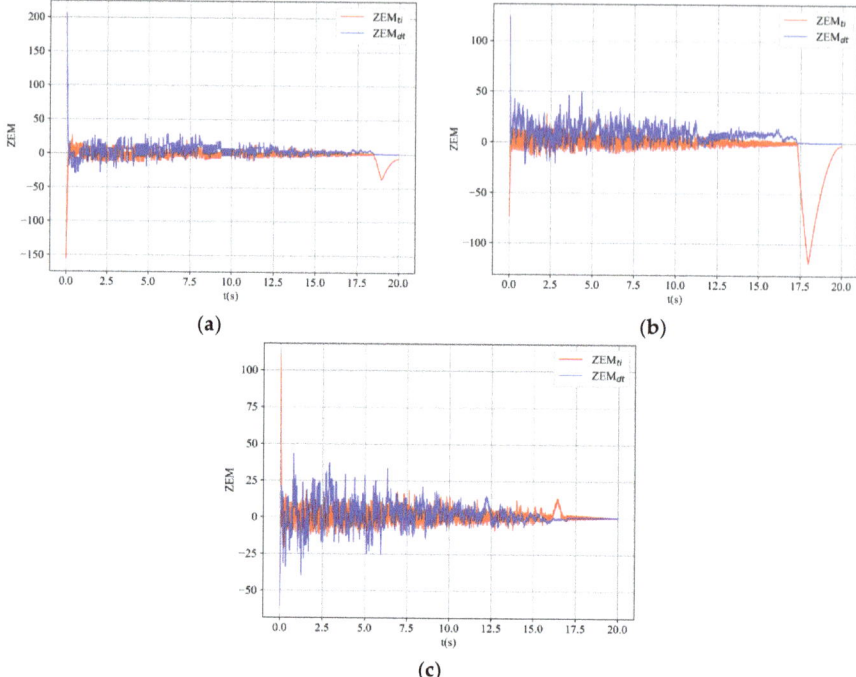

Figure 10. ZEM curve between each spacecraft. (a) dis_{DT} = 5 km, (b) dis_{DT} = 10 km, (c) dis_{DT} = 15 km.

As evident from Figure 11, employing the ICAAI intelligent game algorithm results in the target achieving success rates of no less than approximately 90% when the relative distance to the defender is less than 10 km. However, as the dis_{DT} increases from 10 to 15 km, the success rate of target evasion decreases from 90% to 0%. These simulation results illustrate that a smaller relative distance leads to an increased evasion success rate Additionally, the curve depicting the average miss distance for the target reveals that the miss distance follows a pattern of initially increasing and then decreasing with dis_{DT}. The miss distance reaches its maximum value of approximately 50 m around a relative distance of 5 km. The occurrence of this phenomenon can be attributed to the fact that, when dis_{DT} is less than 5 km, the miss distance increases with the target's evasion time. Moreover, at this point, the interceptor has not had sufficient time to alter its trajectory to intercept the target. Conversely, when dis_{DT} exceeds 5 km, the interceptor has ample time to intercept the target after evading the defender. Consequently, the miss distance decreases with an increasing dis_{DT}.

4.5. Experiment 3: Adaptiveness of the ICAAI Guidance

In the real-world game confrontation process, obtaining the opponent's prior knowledge, such as the maximum acceleration and time constant, is often impractical. To assess the proposed ICAAI guidance method's superior adaptability compared to the OGL method under conditions of unknown opponent knowledge, several comparison conditions were designed and evaluated using the Monte Carlo target shooting method. The adaptive capabilities of both methods were analyzed based on the game results (escape success rate and miss distance) of the target spacecraft employing the two strategies.

While the target utilized OGL guidance, we considered it adopting $\overline{u}_I^{max} = 8$ g, $\overline{\tau}_I = 0.02$ s as the prediction of the prior knowledge of the interceptor, while the actual $u_I^{max} = 6 \sim 10$ g, $\tau = 0.05 \sim 0.002$ s. The simulation results are shown in Figure 12.

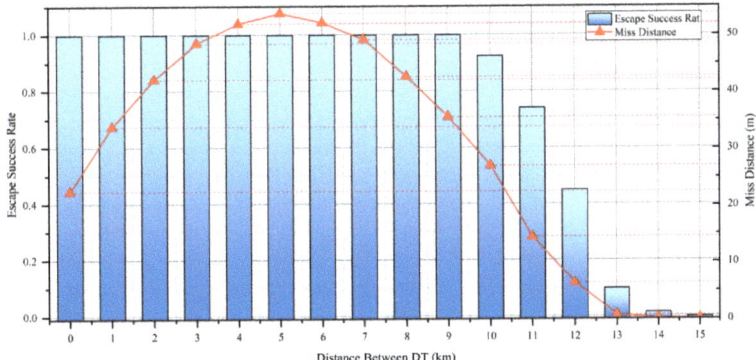

Figure 11. Target game results under different distances between target and defender.

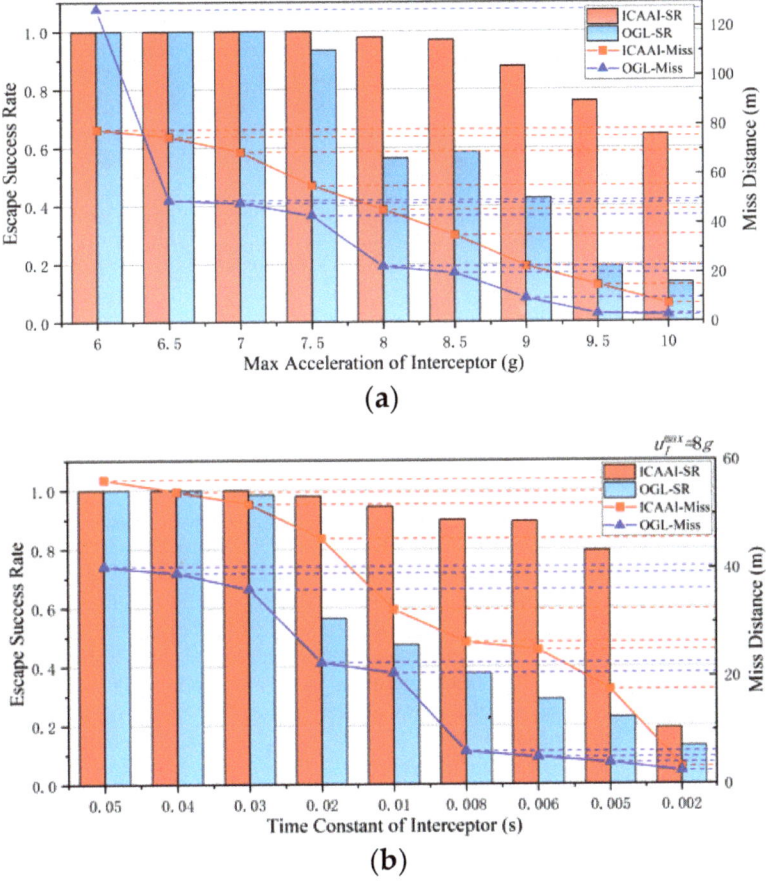

Figure 12. Simulation results in situations without prior knowledge. (**a**) $u_I^{\max} = 6 \sim 10$ g, $\tau = 0.02$ s, (**b**) $u_I^{\max} = 8$ g, $\tau = 0.05 \sim 0.002$ s.

As depicted in Figure 12a, as the interceptor's maneuverability improves, the target's escape ability decreases for both guidance methods. However, it is evident that, when employing the ICAAI guidance, the rate of decline in the target's escape ability is significantly lower compared to the OGL guidance method. Similarly, Figure 12b demonstrates that an increase in the interceptor's response speed yields a similar trend in the target's escape ability as in Figure 12a. Specifically, when accurately estimating the prior knowledge of the target, the escape abilities of both methods are comparable. However, when the prior knowledge error exceeds 25%, the OGL guidance leads to a reduction of over 75% in the target's escape ability, while the ICAAI guidance results in less than a 34% decrease. In conclusion, the proposed ICAAI guidance exhibits superior adaptability compared to the OGL guidance when the interceptor's prior knowledge is unknown.

Remark 2. *As an analytical method, the SOGL is stable but inflexible due to its theoretical framework [46] and stringent assumptions [47]. Correspondingly, the ICAAI control strategies are flexible and can be continuously optimized. The proposed method is independent of the time constant, which means that it performs better with less prior knowledge than the OGL. Furthermore, the adaptability of the proposed method can be improved by considering the tolerance of the maximum interceptor acceleration.*

4.6. Experiment 4: Robustness of the RL-Based Guidance Method

In addition to the unperturbed, fully observable game, the following noisy, partially observable game studies have been analyzed separately in this manuscript. The parameters used to describe the imperfect information model defined in Section 3 are shown in Table 5 The Monte Carlo simulation method is used to obtain the escape success rate and the miss distance of the target using the proposed ICAAI guidance and SOGL guidance under different noise conditions. The results of the Monte Carlo simulation are shown in Figure 13.

Table 5. Parameters of the different imperfect information models.

Measurement Noise	Parameter	Case 1	Case 2	Case 3
LOS	σ_{LOS}(mrad)	0.05	0~0.2	0.05
Velocity	σ_v(m/s)	0.2	0.2	0~0.5
Acceleration	σ_a(m/s)	1~3	2	2

Based on the simulation results of Case 2, it was observed that the OGL method exhibited significant sensitivity to LOS noise. In scenarios without LOS noise, the escape success rate of the proposed ICAAI guidance matched that of the OGL guidance, and, in some cases, the OGL method even achieved a larger miss distance. However, as the LOS noise variance increased to 0.05 mrad, the success rate of the OGL method dropped to approximately 50%. Eventually, at a LOS noise variance of 0.15 mrad, the target was practically unable to escape using the SOGL method, while the ICAAI guidance still maintained an escape success rate of around 80%.

Analyzing the simulation results of Case 1 and Case 3, it was found that due to the presence of LOS noise, the target employing the OGL method exhibited reduced sensitivity to acceleration and velocity noise. Nevertheless, its escape capability remained weaker compared to that of the ICAAI guidance. This could be attributed to the policy network propagating observation information with different weights, leveraging the exploration mechanism of reinforcement learning (RL). Consequently, training the agent in a deterministic environment resulted in a robust guidance policy with strong noise-resistant ability.

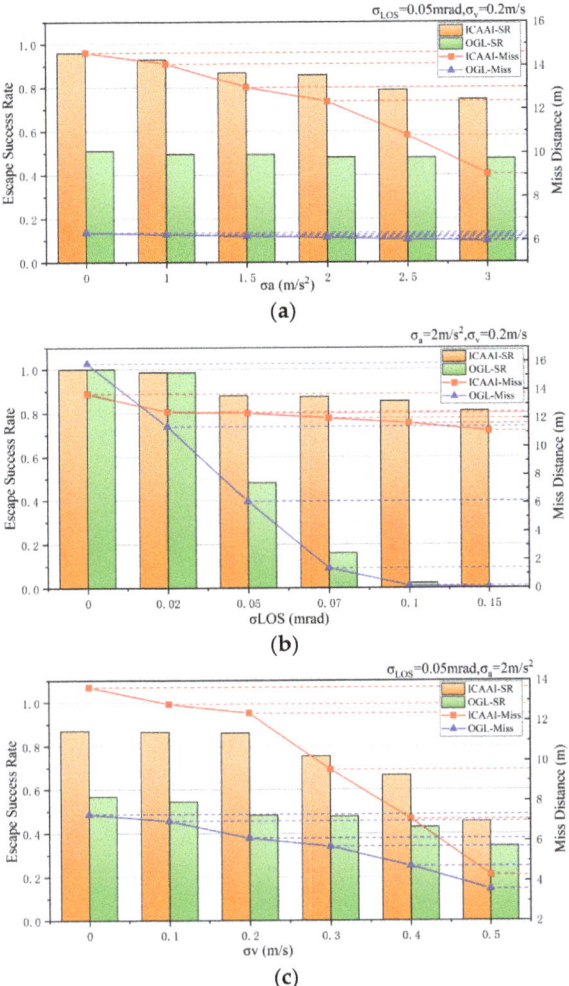

Figure 13. Simulation results in noise-corrupted environment. (**a**) Case 1, (**b**) Case 2, (**c**) Case 3.

5. Conclusions

In this research, we solved the cooperative active defense guidance problem for a target with active defense attempting to evade an interceptor. Based on deep reinforcement learning algorithms, a collaborative guidance strategy termed ICAAI was formulated to enhance active spacecraft defense. Monte Carlo simulations were conducted to empirically substantiate the real-time performance, convergence, adaptiveness, and robustness of the introduced guidance strategy. The conclusions are stated as follows:

(1) In the presence of less prior knowledge and observation noise, the proposed ICAAI guidance strategy is effective in achieving a higher success rate of target evasion by guiding the target to coordinate maneuvers with defensive spacecraft.

(2) Utilizing a heuristic continuous reward function and an adaptive progressive curriculum learning method, we devised the ESC training approach to effectively tackle issues of low convergence efficiency and training process instability in ICAAI.

(3) The ICAAI guidance strategy outperforms the linear–quadratic optimal guidance law (LQOGL) [10] in real-time performance. This framework also achieved an impressive

update frequency of 10^3 Hz, demonstrating substantial potential for onboard applications.

(4) Simulation results confirm ICAAI's effectiveness in reducing the relative distance between interceptor and defender, enabling successful target evasion. In contrast to traditional OGL methods, our approach exhibits enhanced robustness in noisy environments, particularly in mitigating line-of-sight (LOS) noise.

Author Contributions: Conceptualization, W.N., J.L., Z.L., P.L. and H.L.; Methodology, W.N., J.L. Z.L., P.L. and H.L.; Software, W.N.; Validation, W.N.; Investigation, W.N.; Data curation, W.N. Writing—original draft, W.N.; Writing—review & editing, H.L.; Project administration, H.L.; Funding acquisition, J.L. and P.L. All authors have read and agreed to the published version of the manuscript.

Funding: The work described in this paper is supported by the National Natural Science Foundation of China (Grant No. 62003375). The authors fully appreciate their financial supports.

Data Availability Statement: Not applicable.

Conflicts of Interest: The authors declare no conflict of interest.

Nomenclature

a	acceleration, m/s^2
$\mathbf{A}, \mathbf{B}$	state-space model of the linearized equations of motion
H	Hamiltonian
$\mathbf{I}$	identity matrix
$J(\cdot)$	cost function
$\mathbf{L}$	constant vector
L^{-1}	inverse Laplace transform
LOS	light-of-sight
$Q(\cdot)$	reward signal
r	reward signal
s	state defined in Markov decision process
o	observation of the agent
t, t_{go}, t_f	time, time to go, and final time, respectively, s
u	guidance command, m/s^2
V	velocity, m/s
$X-O-Y$	Cartesian reference frame
$\mathbf{x}$	state vector of the linearized equations of motion
y	lateral distance, m
Z	zero-effort-miss, m
α, β, σ	design parameters of the reward function
ϕ	flight path angle, rad
Φ	transition matrix
γ	discount factor
η	killing radius, m
λ	the angle between the corresponding light-of-sight and X-axis, rad
$\lambda(\cdot)$	Lagrange multiplier vector
$\mu(\cdot)$	policy function
ρ	relative distance between the adversaries, m
τ	time constant
ω, ζ	design parameters of the optimal guidance law (OGL)
I, T, D	interceptor, target, and defender, respectively
max	maximum
*	optimal solution

References

1. Ye, D.; Shi, M.; Sun, Z. Satellite proximate pursuit-evasion game with different thrust configurations. *Aerosp. Sci. Technol.* **2020**, *99*, 105715. [CrossRef]
2. Boyell, R.L. Defending a moving target against missile or torpedo attack. *IEEE Trans. Aerosp. Electron. Syst.* **1976**, *AES-12*, 522–526. [CrossRef]

6. Rusnak, I. Guidance laws in defense against missile attack. In Proceedings of the 2008 IEEE 25th Convention of Electrical and Electronics Engineers in Israel, Eilat, Israel, 3–5 December 2008; pp. 090–094.
7. Rusnak, I. The lady, the bandits and the body guards—A two team dynamic game. *IFAC Proc. Vol.* **2005**, *38*, 441–446. [CrossRef]
8. Shalumov, V. Optimal cooperative guidance laws in a multiagent target–missile–defender engagement. *J. Guid. Control Dyn.* **2019**, *42*, 1993–2006. [CrossRef]
9. Weiss, M.; Shima, T.; Castaneda, D.; Rusnak, I. Combined and cooperative minimum-effort guidance algorithms in an active aircraft defense scenario. *J. Guid. Control Dyn.* **2017**, *40*, 1241–1254. [CrossRef]
10. Weiss, M.; Shima, T.; Castaneda, D.; Rusnak, I. Minimum effort intercept and evasion guidance algorithms for active aircraft defense. *J. Guid. Control Dyn.* **2016**, *39*, 2297–2311. [CrossRef]
11. Shima, T. Optimal cooperative pursuit and evasion strategies against a homing missile. *J. Guid. Control. Dyn.* **2011**, *34*, 414–425. [CrossRef]
12. Perelman, A.; Shima, T.; Rusnak, I. Cooperative differential games strategies for active aircraft protection from a homing missile. *J. Guid. Control Dyn.* **2011**, *34*, 761–773. [CrossRef]
13. Liang, H.; Wang, J.; Wang, Y.; Wang, L.; Liu, P. Optimal guidance against active defense ballistic missiles via differential game strategies. *Chin. J. Aeronaut.* **2020**, *33*, 978–989. [CrossRef]
14. Anderson, G.M. Comparison of optimal control and differential game intercept missile guidance laws. *J. Guid. Control* **1981**, *4*, 109–115. [CrossRef]
15. Dong, J.; Zhang, X.; Jia, X. Strategies of pursuit-evasion game based on improved potential field and differential game theory for mobile robots. In Proceedings of the 2012 Second International Conference on Instrumentation, Measurement, Computer, Communication and Control, Harbin, China, 8–10 December 2012; pp. 1452–1456.
16. Li, Z.; Wu, J.; Wu, Y.; Zheng, Y.; Li, M.; Liang, H. Real-time Guidance Strategy for Active Defense Aircraft via Deep Reinforcement Learning. In Proceedings of the NAECON 2021-IEEE National Aerospace and Electronics Conference, Dayton, OH, USA, 16–19 August 2021; pp. 177–183.
17. Liang, H.; Li, Z.; Wu, J.; Zheng, Y.; Chu, H.; Wang, J. Optimal Guidance Laws for a Hypersonic Multiplayer Pursuit-Evasion Game Based on a Differential Game Strategy. *Aerospace* **2022**, *9*, 97. [CrossRef]
18. Liu, F.; Dong, X.; Li, Q.; Ren, Z. Cooperative differential games guidance laws for multiple attackers against an active defense target. *Chin. J. Aeronaut.* **2022**, *35*, 374–389. [CrossRef]
19. Weintraub, I.E.; Cobb, R.G.; Baker, W.; Pachter, M. Direct methods comparison for the active target defense scenario. In Proceedings of the AIAA Scitech 2020 Forum, Orlando, FL, USA, 6–10 January 2020; p. 0612.
20. Shalumov, V. Cooperative online guide-launch-guide policy in a target-missile-defender engagement using deep reinforcement learning. *Aerosp. Sci. Technol.* **2020**, *104*, 105996. [CrossRef]
21. Liang, H.; Wang, J.; Liu, J.; Liu, P. Guidance strategies for interceptor against active defense spacecraft in two-on-two engagement. *Aerosp. Sci. Technol.* **2020**, *96*, 105529. [CrossRef]
22. Salmon, J.L.; Willey, L.C.; Casbeer, D.; Garcia, E.; Moll, A.V. Single pursuer and two cooperative evaders in the border defense differential game. *J. Aerosp. Inf. Syst.* **2020**, *17*, 229–239. [CrossRef]
23. Harel, M.; Moshaiov, A.; Alkaher, D. Rationalizable strategies for the navigator–target–missile game. *J. Guid. Control Dyn.* **2020**, *43*, 1129–1142. [CrossRef]
24. Miljković, Z.; Mitić, M.; Lazarević, M.; Babić, B. Neural network reinforcement learning for visual control of robot manipulators. *Expert Syst. Appl.* **2013**, *40*, 1721–1736. [CrossRef]
25. Ye, D.; Chen, G.; Zhang, W.; Chen, S.; Yuan, B.; Liu, B.; Chen, J.; Liu, Z.; Qiu, F.; Yu, H. Towards playing full moba games with deep reinforcement learning. *arXiv* **2020**, arXiv:2011.12692.
26. Shalev-Shwartz, S.; Shammah, S.; Shashua, A. Safe, multi-agent, reinforcement learning for autonomous driving. *arXiv* **2016**, arXiv:1610.03295.
27. Zhu, Y.; Mottaghi, R.; Kolve, E.; Lim, J.J.; Gupta, A.; Fei-Fei, L.; Farhadi, A. Target-driven visual navigation in indoor scenes using deep reinforcement learning. In Proceedings of the 2017 IEEE International Conference on Robotics and Automation (ICRA), Singapore, 29 May 2017–3 June 2017; pp. 3357–3364.
28. Mnih, V.; Kavukcuoglu, K.; Silver, D.; Rusu, A.A.; Veness, J.; Bellemare, M.G.; Graves, A.; Riedmiller, M.; Fidjeland, A.K.; Ostrovski, G.; et al. Human-level control through deep reinforcement learning. *Nature* **2015**, *518*, 529–533. [CrossRef] [PubMed]
29. Gaudeta, B.; Furfaroa, R.; Linares, R. Reinforcement meta-learning for angle-only intercept guidance of maneuvering targets. In Proceedings of the AIAA Scitech 2020 Forum AIAA 2020, Orlando, FL, USA, 6–10 January 2020; Volume 609.
30. Gaudet, B.; Linares, R.; Furfaro, R. Adaptive guidance and integrated navigation with reinforcement meta-learning. *Acta Astronaut.* **2020**, *169*, 180–190. [CrossRef]
31. Lau, M.; Steffens, M.J.; Mavris, D.N. Closed-loop control in active target defense using machine learning. In Proceedings of the AIAA Scitech 2019 Forum, San Diego, CA, USA, 7–11 January 2019; p. 0143.
32. Zhang, G.; Chang, T.; Wang, W.; Zhang, W. Hybrid threshold event-triggered control for sail-assisted USV via the nonlinear modified LVS guidance. *Ocean Eng.* **2023**, *276*, 114160. [CrossRef]
33. Li, J.; Zhang, G.; Shan, Q.; Zhang, W. A novel cooperative design for USV–UAV systems: 3D mapping guidance and adaptive fuzzy control. *IEEE Trans. Control Netw. Syst.* **2022**, *10*, 564–574. [CrossRef]

31. Ainsworth, M.; Shin, Y. Plateau phenomenon in gradient descent training of RELU networks: Explanation, quantification, and avoidance. *SIAM J. Sci. Comput.* **2021**, *43*, A3438–A3468. [CrossRef]
32. Fujimoto, S.; Hoof, H.; Meger, D. Addressing function approximation error in actor-critic methods. In *PMLR, Proceedings of Machine Learning Research, Proceedings of the 35th International Conference on Machine Learning, Stockholm, Sweden, 10–15 July 2018*; PMLR: New York, NY, USA, 2018; Volume 80, pp. 1587–1596.
33. Schulman, J.; Wolski, F.; Dhariwal, P.; Radford, A.; Klimov, O. Proximal policy optimization algorithms. *arXiv* **2017**, arXiv:1707.06347.
34. Babaeizadeh, M.; Frosio, I.; Tyree, S.; Clemons, J.; Kautz, J. Reinforcement learning through asynchronous advantage actor-critic on a gpu. *arXiv* **2016**, arXiv:1611.06256.
35. Casas, N. Deep deterministic policy gradient for urban traffic light control. *arXiv* **2017**, arXiv:1703.09035.
36. Silver, D.; Lever, G.; Heess, N.; Degris, T.; Wierstra, D.; Riedmiller, M. Deterministic Policy Gradient Algorithms. In *PMLR, Proceedings of Machine Learning Research, Proceedings of the 31st International Conference on Machine Learning, Beijing China, 21–26 June 2014*; PMLR: New York, NY, USA, 2014; Volume 32, pp. 387–395.
37. Fan, J.; Wang, Z.; Xie, Y.; Yang, Z. A Theoretical Analysis of Deep Q-Learning. In *PMLR, Proceedings of Machine Learning Research, Proceedings of the 2nd Conference on Learning for Dynamics and Control, Online, 10–11 June 2020*; PMLR: New York, NY, USA, 2020; Volume 120, pp. 486–489.
38. Hasselt, H. Double Q-learning. In *Advances in Neural Information Processing Systems*; Curran Associates Inc.: New York, NY, USA, 2010; Volume 23.
39. Gullapalli, V.; Barto, A.G. Shaping as a method for accelerating reinforcement learning. In Proceedings of the 1992 IEEE International Symposium on Intelligent Control, Glasgow, UK, 11–13 August 1992; pp. 554–559.
40. Bengio, Y.; Louradour, J.; Collobert, R.; Weston, J. Curriculum learning. In Proceedings of the 26th Annual International Conference on Machine Learning, Montreal, QC, Canada, 14–18 June 2009; pp. 41–48.
41. Krueger, K.A.; Dayan, P. Flexible shaping: How learning in small steps helps. *Cognition* **2009**, *110*, 380–394. [CrossRef]
42. Ng, A.Y.; Harada, D.; Russell, S. Policy invariance under reward transformations: Theory and application to reward shaping. *LCML* **1999**, *99*, 278–287.
43. Randløv, J.; Alstrøm, P. Learning to Drive a Bicycle Using Reinforcement Learning and Shaping. *ICML* **1998**, *98*, 463–471.
44. Wiewiora, E. Potential-based shaping and Q-value initialization are equivalent. *J. Artif. Intell. Res.* **2003**, *19*, 205–208. [CrossRef]
45. Qi, N.; Sun, Q.; Zhao, J. Evasion and pursuit guidance law against defended target. *Chin. J. Aeronaut.* **2017**, *30*, 1958–1973. [CrossRef]
46. Ho, Y.; Bryson, A.; Baron, S. Differential games and optimal pursuit-evasion strategies. *IEEE Trans. Autom. Control* **1965**, *10*, 385–389. [CrossRef]
47. Shinar, J.; Steinberg, D. Analysis of Optimal Evasive Maneuvers Based on a Linearized Two-Dimensional Kinematic Model. *J. Aircr.* **1977**, *14*, 795–802. [CrossRef]

Disclaimer/Publisher's Note: The statements, opinions and data contained in all publications are solely those of the individual author(s) and contributor(s) and not of MDPI and/or the editor(s). MDPI and/or the editor(s) disclaim responsibility for any injury to people or property resulting from any ideas, methods, instructions or products referred to in the content.

A Reentry Trajectory Planning Algorithm via Pseudo-Spectral Convexification and Method of Multipliers

Haizhao Liang [1,2], Yunhao Luo [1], Haohui Che [1], Jingxian Zhu [1] and Jianying Wang [1,*]

1. School of Aeronautics and Astronautics, Shenzhen Campus of Sun Yat-sen University, Shenzhen 518107, China; lianghch5@mail.sysu.edu.cn (H.L.); luoyh73@mail2.sysu.edu.cn (Y.L.); chehh@mail2.sysu.edu.cn (H.C.); zhujx26@mail2.sysu.edu.cn (J.Z.)
2. Shenzhen Key Laboratory of Intelligent Microsatellite Constellation, Shenzhen 518107, China
* Correspondence: wangjiany@mail.sysu.edu.cn

Abstract: The reentry trajectory planning problem of hypersonic vehicles is generally a continuous and nonconvex optimization problem, and it constitutes a critical challenge within the field of aerospace engineering. In this paper, an improved sequential convexification algorithm is proposed to solve it and achieve online trajectory planning. In the proposed algorithm, the Chebyshev pseudo-spectral method with high-accuracy approximation performance is first employed to discretize the continuous dynamic equations. Subsequently, based on the multipliers and linearization methods, the original nonconvex trajectory planning problem is transformed into a series of relaxed convex subproblems in the form of an augmented Lagrange function. Then, the interior point method is utilized to iteratively solve the relaxed convex subproblem until the expected convergence precision is achieved. The convex-optimization-based and multipliers methods guarantee the promotion of fast convergence precision, making it suitable for online trajectory planning applications. Finally, numerical simulations are conducted to verify the performance of the proposed algorithm. The simulation results show that the algorithm possesses better convergence performance, and the solution time can reach the level of seconds, which is more than 97% less than nonlinear programming algorithms, such as the sequential quadratic programming algorithm.

Keywords: reentry trajectory planning; improved sequential convexification; hypersonic vehicle; pseudo-spectral method; method of multipliers

MSC: 49M37

1. Introduction

Hypersonic vehicles generally refer to near-space vehicles with flight speeds greater than Mach 5. They possess the advantages of strong maneuverability, flexible trajectory, they are difficult to intercept, and so on, and they have been increasingly valued by the major space powers due to their high flight speeds and vast airspace coverage. Among the related technologies, trajectory planning technology can provide important support for performance analysis regarding flight range, maneuverability, and ballistic characteristics. The hypersonic vehicle trajectory planning problem typically involves solving a nonlinear optimal control problem with various state and control constraints, including boundary conditions, no-fly-zone constraints, path constraints, etc. [1–3].

In this regard, many scholars and engineers have carried out a series of in-depth studies, and the proposed trajectory planning algorithms generally include indirect and direct methods. The indirect methods transform the trajectory planning problem into a Hamiltonian boundary value problem based on the Pontryagin maximum principle and solves it by employing the gradient method and other algorithms. On the other hand, many scholars have proposed and developed the collocation method, pseudo-spectral method, and other methods [4–7], demonstrating the advantages of direct methods in solving trajectory optimization problems. However, the direct methods still have some shortcomings under the requirement of rapid trajectory planning, such as high sensitivity to initial guess value and uncertain solving time and convergence, which restrict their efficient solving ability [8].

In recent years, convex-optimization-based methods have effectively met the demand for efficient solutions and attracted more and more attention in terms of spacecraft trajectory optimization. When a problem can be formulated within a convex optimization framework, its complexity is low and can be reliably solved to global optimality in the polynomial time by the primal-dual interior point method. The upper bound of the number of iterations required for convergence is also determined. Moreover, the primal-dual interior point method can be adopted to solve the convex problem without the initial guess value. Motivated by these preponderances, the convex-optimization-based methods have been applied to different aerospace problems, such as planetary reentry trajectory optimization [9], ascent trajectory optimization [10,11], Mars-landing trajectory planning [12,13], low-thrust orbit transfer [14], spacecraft rendezvous and proximity operations [15,16], and trajectory planning for satellite cluster reconfigurations [17].

Most aerospace problems are limited by nonlinear, nonconvex dynamics and path constraints, so they cannot be solved directly under the convex optimization framework. Therefore, convexification technologies that make the approximate error as small as possible are a significant research direction [18]. Among them, the two mainstream convexification methods include lossless convexification and sequential convexification methods. In Refs. [12,13], Ackimese et al. employed the lossless convexification method to solve the Mars-landing trajectory planning problem by replacing nonconvex constraints with relaxed convex constraints without a loss of accuracy. But the lossless convexification method is only suitable for a few constraints with particular forms, which limits its wide application. On the other hand, the sequential convexification method can conduct highly nonlinear complex problems. The basic idea is to obtain a series of convex subproblems by approximating the nonlinear terms and then solve them iteratively until it converges to the expected precision [19,20].

However, the sequential convexification method can be further improved in terms of discretization and accelerating convergence. Firstly, the traditional trapezoidal discrete method [3,19,20] is often chosen to discretize the continuous optimization problem, which leads to a large deviation between the approximate model used in the solution procedure and the actual one. To obtain a precise-enough solution, the equidistant discrete nodes should be sufficiently numerous. Nevertheless, it also results in a dramatic increase in the number of optimization variables and takes lots of time to solve. In contrast, the pseudo-spectral discretization methods offer higher accuracy under the same number of discrete nodes and have been widely employed for solving the optimal control problem [5,8]. Among the pseudo-spectral methods, the Chebyshev pseudo-spectral discretization method [21,22] is a special category, which can minimize the Runge phenomenon and supply the best polynomial approximation under the minimax norm. On the other hand, "artificial infeasibility" [23,24] caused by the convexification errors occurs when the original problem is feasible but the convex subproblem is not. Researchers address this issue by introducing slack variables to relax the feasible domain limited by various constraints, with large constant penalty parameters added to penalize these slack variables. As a result, the "artificial infeasible" gradually disappears and the slack variables tend towards zero as the iterative solution converges. Unfortunately, the fixed penalty parameters could cause the solution to converge to a stagnation point of the penalty problem rather than the original problem, according to Ref. [25].

To address the aforementioned issues, an improved sequential convexification algorithm is proposed to improve the performance of solving the trajectory planning problem of hypersonic vehicles in this paper. Firstly, the Chebyshev pseudo-spectral method with higher approximation precision is employed to discretize the continuous optimal control problem. And the flight terminal time is designed as an optimization variable, so that it can be applied to the optimal control problem with a fixed initial state and free terminal state. Then, by introducing automatically updated penalty parameters and Lagrange multipliers, the relaxed convex subproblem in the form of the augmented Lagrange function is constructed to improve the convergence and computational properties based on the method of multipliers (or the augmented Lagrange method). And it is solved iteratively by the interior point method in the framework of the improved sequential convexification algorithm.

The rest of this paper is organized as follows. In Section 2, the reentry trajectory optimization problem is formulated. In Section 3, the improved sequential convexification algorithm is detailed. In Section 4, numerical simulations are presented to verify the performance of the algorithm. Finally, the conclusion and discussion are provided in Section 5.

2. Problem Formulation

2.1. Reentry Dynamics

In view of the dynamics characteristics of the reentry flight of hypersonic vehicles, the dimensionless three-degrees-of-freedom augmented dynamics model is established in the half-velocity coordinate system without considering Earth's flatness and rotation:

$$\begin{cases} \dot{r} = v \sin \gamma \\ \dot{\theta} = v \cos \gamma \sin \psi / (r \cos \phi) \\ \dot{\phi} = v \cos \gamma \cos \psi / r \\ \dot{v} = -D - \sin \gamma / r^2 \\ \dot{\gamma} = L \cos \sigma / v + (v^2 - 1/r) \cos \gamma / (vr) \\ \dot{\psi} = L \sin \sigma / (v \cos \gamma) + v \cos \gamma \sin \psi \tan \phi / r \\ \dot{\sigma} = u \end{cases} \quad (1)$$

where r is the radial distance from the Earth center to the vehicle, θ and ϕ are the longitude and latitude, respectively, v is the velocity, γ is the flight path angle, ψ is the heading angle, and σ is the bank angle. The above variables are defined as system state variables, i.e., $x = [r, v, \theta, \phi, \gamma, \psi, \sigma]^T$, and the control variable is the bank angle rate $u \in R$. The variables r, v and time are scaled by R_0, $\sqrt{R_0 g_0}$ and $\sqrt{R_0/g_0}$, respectively, where $R_0 = 6371$ km is the Earth radius and $g_0 = 9.8$ m/s^2 is the gravitational acceleration at sea level. Dimensionless lift and drag accelerations L and D are scaled by g_0 and calculated as:

$$\begin{cases} L = \frac{R_0 \rho v^2 S C_L}{2m} \\ D = \frac{R_0 \rho v^2 S C_D}{2m} \end{cases} \quad (2)$$

where m is the vehicle mass, S is the reference area of the vehicle, and ρ is the atmospheric density, $\rho = \rho_0 e^{-H/H_s}$, where $H = rR_0 - R_0$ is the height, H_s is the atmospheric density scale height, and ρ_0 is the atmospheric density at sea level. Moreover, C_L and C_D are lift and drag coefficients, respectively. For reference to the aerodynamic parameters of the vehicle in [9], the aerodynamic coefficient can be expressed as:

$$\begin{cases} C_L = -0.041065 + 0.016292\alpha + 0.0002602\alpha^2 \\ C_D = 0.080505 - 0.03026 C_L + 0.86495 C_L^2 \end{cases} \quad (3)$$

in which the angle of attack α is in degree and expressed as a function of velocity. The angle-of-attack velocity profile is preset as follows:

$$\alpha = \begin{cases} 40, & v\sqrt{R_0 g_0} > 4570 \text{ m/s} \\ 40 - 0.20705 \frac{(v\sqrt{R_0 g_0} - 4570)^2}{340^2}, & \text{otherwise} \end{cases} \quad (4)$$

In general, the design of the bank angle is the main means to change the trajectory of the vehicle when the angle of attack is preset. However, the bank angle rate is adopted here as the new control variable for the following reasons. Firstly, boundary constraints are applied to make the bank angle profile smoother. Secondly, the current system dynamics are transformed into affine control form, which expedite subsequent convexification operations and alleviate the high-frequency fluctuations in the bank angle [9].

2.2. Constraint Conditions

The trajectory planning problem of hypersonic vehicles belongs to a class of optimal control problems, and its purpose is to determine the optimal control variables and optimize the performance index function under various constraints.

Generally, the vehicle must meet different constraints, such as heat rate, dynamic pressure, load, and control margin, during the ultra-high-speed flight to ensure the safety of the vehicle structure, thermal protection, guidance, and control systems. To meet the mission requirements, the initial conditions and terminal states of the vehicle also need to be limited. In addition, the vehicle should avoid certain no-fly zones due to radar detection, geopolitics, or other considerations. In summary, all of these constraint conditions can be divided into two categories: equality constraints and inequality constraints.

First of all, the equality constraints mainly include dynamic equations and initial and terminal constraints. Here, the system dynamics Equation (1) is abstractly expressed as

$$\dot{x} = f(x, u, t) = f_0(x, t) + Bu, \tag{5}$$

where t is the system time variable, and $f(\cdot, \cdot, \cdot)$ is the right function of the dynamic equations. And $f_0(x,t) = \begin{cases} v \sin \gamma \\ v \cos \gamma \sin \psi / (r \cos \phi) \\ v \cos \gamma \cos \psi / r \\ -D - \sin \gamma / r^2 \\ L \cos \sigma / v + (v^2 - 1/r) \cos \gamma / (vr) \\ L \sin \sigma / (v \cos \gamma) + v \cos \gamma \sin \psi \tan \phi / r \\ 0 \end{cases}$, $B = [0, 0, 0, 0, 0, 0, 1]^T$.

The initial and terminal constraints are determined by the flight mission, which contains the requirements of the reentry start point and the target point, and they can be represented as follows:

$$\begin{aligned} \Phi(x(t_0), x_0) &= 0 \\ \Psi(x(t_f), x_f) &= 0 \end{aligned}, \tag{6}$$

where t_0 and t_f are the initial and terminal time, respectively. $\Phi(\cdot, \cdot)$ and $\Psi(\cdot, \cdot)$ are the initial and terminal state constraints, respectively.

Then, inequality constraints can be divided into the following three categories. First, the path constraints, including the maximal heat rate, dynamic pressure, and load, are expressed as:

$$p(r,v) = \begin{bmatrix} \dot{Q} - \dot{Q}_{\max} \\ q - q_{\max} \\ n - n_{\max} \end{bmatrix} = \begin{bmatrix} K_Q \rho^{0.5} (v \sqrt{R_0 g_0})^{3.15} - \dot{Q}_{\max} \\ 0.5 \rho (v \sqrt{R_0 g_0})^2 - q_{\max} \\ \sqrt{L^2 + D^2} - n_{\max} \end{bmatrix} \leq 0, \tag{7}$$

where $\dot{Q}_{\max}$, $q_{\max}$ and $n_{\max}$ are the corresponding maximum values, respectively.

Second, the bounded constraints about the state and control variables are given by:

$$\begin{aligned} x_{\min} &\leq x \leq x_{\max} \\ u_{\min} &\leq u \leq u_{\max} \end{aligned}, \tag{8}$$

where $\cdot_{\min}$ and $\cdot_{\max}$ are the lower and upper bound values, respectively.

The last class of inequality constraints is the no-fly-zone (NFZ) constraint. In general, an NFZ is modeled as a circular exclusion zone in the horizontal place with infinite height and limits the longitude and latitude of the flight trajectory by

$$(\theta - \theta_c)^2 + (\phi - \phi_c)^2 \geq d^2, \tag{9}$$

where θ_c and ϕ_c are the longitude and latitude at the center of the NFZ, and d is the radius.

2.3. Optimal Control Problem

In the trajectory optimization problem, the performance indexes can be reasonably selected according to the different flight tasks and design requirements. Common performance indexes include minimum flight time, maximum range, minimum heat load, etc. In this paper, the minimum flight time is chosen as the performance index, considering that the vehicle needs to reach the anticipative target point quickly. Henceforth, the original reentry trajectory optimization problem can be formulated as a highly constrained optimal control problem:

$$P_0: \quad \min \quad J = \int_0^{t_f} 1 \, dt$$
$$\text{s.t. Eq. (5), (6), (7), (8), (9)}.$$

3. Improved Sequential Convexification Algorithm

In this section, an augmented Lagrange-based Chebyshev pseudo-spectral form improved sequential convexification (AL-CP-ISC) algorithm is proposed to solve the above continuous and nonlinear reentry trajectory optimization problem. The advantages of the pseudo-spectral method with high discrete accuracy and the method of multipliers with good convergence performance and stable numerical computation are synthesized to empower the AL-CP-ISC algorithm. Firstly, the discretization and convexification process of the vehicle's nonlinear system are given on the strength of the Chebyshev pseudo-spectral method and first-order Taylor expansion. Subsequently, the original problem is transformed into a series of relaxed convex subproblems by introducing the slack variables, penalty parameters, and Lagrange multipliers. Finally, an algorithm solution procedure is presented.

3.1. Discretization and Convexification

The Chebyshev pseudo-spectral discretization method with unique time-domain mapping is adopted to discretize the continuous reentry trajectory optimization problem P_0 with the free terminal time. In the Chebyshev pseudo-spectral discretization method, the domain of the Chebyshev–Gauss–Lobatto (CGL) points is $\tau \in [-1, 1]$, but the flight time interval is $t \in [t_0, t_f]$ in the practical problem. Hence, the time variable is transformed into

$$\tau = \frac{2t}{t_f - t_0} - \frac{t_f + t_0}{t_f - t_0}. \tag{10}$$

The CGL points are unevenly distributed on the interval $[-1, 1]$:

$$\tau_k = \cos\left(\frac{\pi k}{N}\right) \quad k = 0, \ldots, N. \tag{11}$$

Taking the real state and control variables at $N + 1$ nodes above, the Lagrange interpolation polynomials are constructed, respectively, as approximations of continuous state and control variables. The approximate expressions of the real state variable x and the control variable u are

$$x(\tau) \approx x^N(\tau) = \sum_{j=0}^{N} x_j \phi_j(\tau)$$
$$u(\tau) \approx u^N(\tau) = \sum_{j=0}^{N} u_j \phi_j(\tau)$$
$$\tag{12}$$

where $\phi_j(\tau)$ is the Nth Lagrange interpolation basis function

$$\phi_j(\tau) = \frac{(-1)^{j+1}}{N^2 c_j} \frac{(1-\tau^2)\dot{T}_N(\tau)}{\tau - \tau_j}, \tag{13}$$

where $T_N(\tau)$ is the $N\text{th}-\text{order}$ Chebyshev polynomial and

$$c_j = \begin{cases} 2, j = 0, N \\ 1, 1 \le j \le N-1 \end{cases}. \tag{14}$$

Deriving the approximate expression for the state variables $x(\tau)$ yields:

$$\dot{x}(\tau_k) \approx \dot{x}^N(\tau_k) = \sum_{j=0}^{N} x_j \dot{\phi}_j(\tau_k) = \sum_{j=0}^{N} D_{kj} x_j \tag{15}$$

where D_{kj} is the kth row and j^{th} column element of the differential matrix $D^{(N+1)\times(N+1)}$ of the Chebyshev pseudo-spectral discretization method, and the calculation of each element's value of the matrix D is shown in Ref. [21].

The derivatives of the state variables in the dynamic equations are replaced using the right-hand term in Equation (15) and discretized at the nodes, so that the original differential dynamic equation constraints are transformed into discrete algebraic constraints:

$$2\sum_{j=0}^{N} D_{kj} x_j + (t_0 - t_f^k) f(x_k, u_k, t_k) = 2\sum_{j=0}^{N} D_{kj} x_j + (t_0 - t_f^k) f_0(x_k, t_k) + (t_0 - t_f^k) B u_k = 0 \quad k = 0, 1, \cdots, N. \tag{16}$$

Next, the above nonlinear algebraic constraints are linearized based on first-order Taylor expansion as follows:

$$2\sum_{j=0}^{N} D_{kj} x_j + A(x^*, t_f^*) x + (t_0 - t_f^*) B u + [T(x^*, t_f^*) - B u^*] t_f + C = 0, \tag{17}$$

where x^*, u^* and t_f^* are the reference values of the optimization variables $[x, u, t_f]$, respectively, and

$$A(x^*, t_f^*) = \frac{\partial[(t_0 - t_f) f_0(x, t)]}{\partial x}\Big|_{x=x^*, t_f=t_f^*} = (t_0 - t_f^*) \frac{\partial f_0(x, t)}{\partial x}\Big|_{x=x^*}, \tag{18}$$

$$(t_0 - t_f^*) B = [0, 0, 0, 0, 0, 0, (t_0 - t_f^*)]^T, \tag{19}$$

$$T(x^*, t_f^*) = \frac{\partial[(t_0 - t_f) f_0(x, t)]}{\partial t_f}\Big|_{x=x^*, t_f=t_f^*} = -f_0(x, t)|_{x=x^*}, \tag{20}$$

$$C = (t_0 - t_f^*) f_0^* - A(x^*, t_f^*) x^* - T(x^*, t_f^*) t_f^* + B u^* t_f^*. \tag{21}$$

One can see Appendix A for more details on the matrix A. Similarly, the nonlinear path constraints are given by:

$$p(x) = \begin{bmatrix} \dot{Q}(x) - \dot{Q}_{max} \\ q(x) - q_{max} \\ n(x) - n_{max} \end{bmatrix} \approx \begin{bmatrix} \dot{Q}(x^*) + \nabla \dot{Q}(x^*)(x - x^*) - \dot{Q}_{max} \\ q(x^*) + \nabla q(x^*)(x - x^*) - q_{max} \\ n(x^*) + \nabla n(x^*)(x - x^*) - n_{max} \end{bmatrix} \le 0, \tag{22}$$

where

$$\nabla \dot{Q}(x^*) = [\tfrac{\partial \dot{Q}}{\partial r}, \tfrac{\partial \dot{Q}}{\partial v}] = [-0.5k_Q(v\sqrt{R_0 g_0})^{3.15}\tfrac{\sqrt{\rho}}{H_s}, 3.15k_Q(\sqrt{R_0 g_0})^{3.15} v^{2.15}\sqrt{\rho}]|_{x=x^*}$$
$$\nabla q(x^*) = [\tfrac{\partial q}{\partial r}, \tfrac{\partial q}{\partial v}] = [-0.5(v\sqrt{R_0 g_0})^2 \tfrac{R_0 \rho}{H_s}, (\sqrt{R_0 g_0})^2 \rho v]|_{x=x^*} \qquad (23)$$
$$\nabla n(x^*) = [\tfrac{\partial n}{\partial r}, \tfrac{\partial n}{\partial v}] = [-0.5 R_0^2 S\sqrt{C_L^2 + C_D^2}\tfrac{\rho v^2}{m H_s}, R_0 S\sqrt{C_L^2 + C_D^2}\tfrac{\rho v}{m}]|_{x=x^*}$$

The NFZ constraint is a nonconvex function and also needs to be convexified by the first-order Taylor expansion, as shown below:

$$2(\theta^* - \theta_c)\theta + 2(\phi^* - \phi_c)\phi \geq d^2 + \tilde{d}, \qquad (24)$$

where

$$\tilde{d} = -(\theta^* - \theta_c)^2 - (\phi^* - \phi_c)^2 + 2(\theta^* - \theta_c)\theta^* + 2(\phi^* - \phi_c)\phi^*. \qquad (25)$$

Finally, to place limits on the deviation of the state variables between the linearized and original system, a trust region constraint is introduced, so as to reduce the linearization error and improve the convergence of the sequential linear approximation, denoted as

$$\|x - x^*\| \leq \delta, \qquad (26)$$

where $\delta \in \mathbb{R}^7$ is a constant vector, and the inequality is expressed in components. This is a second-order conic constraint that is, itself, convex. Adding the trust region constraint is necessary to guarantee the linearized constraints to legitimately approximate the original constraints. Meanwhile, proposition 2 in Ref. [9] theoretically explains that a feasible solution of the linearized problem satisfies the linearized path constraints, and it also satisfies the original path constraints.

Up to now, all the continuous and nonconvex functions in the trajectory optimization problem have been discretized and convexified, and the problem P_0 is converted into a discretized convex problem:

$$DCP_0: \quad \min \quad J = \int_0^{T_f} 1\, dt$$
$$\text{s.t. Eq. (6), (8), (17), (22), (24), (26)}.$$

3.2. Problem Transformation

Combining the discretization and convexification processes, the original problem is transformed into an augmented Lagrange formal convex problem. The AL-CP-ISC algorithm can profit from this conversion, resulting in convergence-rate promotion and numerical difficulty avoidance according to the following analysis [26–28].

The linearized error causes the feasible region of the original problem to shrink to a certain extent in the process of convexification, resulting in the "artificial infeasible" situation in which the original problem is feasible but the linearized problem is not feasible. Hereon, slack variables are introduced to relax the "hard constraints", such as dynamic equations, path constraints, and terminal conditions, to compensate for the linearization errors [8,18]. In the meantime, penalty parameters are introduced to punish the slack variables. When the iterative solution converges, the slack variables also gradually approach zero due to the penalty imposed. However, the penalty parameters are usually selected as a larger constant value in many studies of spacecraft trajectory optimization. Unfortunately, the constant penalty parameters can cause the iterative solution to converge to the stationary point of the penalty problem instead of the original problem [25].

Therefore, to resolve this matter, the slack variables are penalized by penalty parameters, which are automatically updated incrementally. Meanwhile, to prevent the condition number of the Hessian matrix from getting worse and worse when the penalty parameters are updated to infinity, which leads to numerical difficulties in the algorithm, Lagrange multipliers are introduced to transform the penalty function problem into the augmented

Lagrange function one. According to the augmented Lagrange function method, by embedding the multiplier update mechanism, the satisfaction degree of constraints and the optimization of the objective function can be considered in each iteration, thus reducing the number of iterations required to achieve the same precision. Additionally, the augmented Lagrange function method can also effectively control the condition number of the Hessian matrix through its unique construction, which avoids the instability of numerical computation. In short, the iterative solution of the algorithm can converge to the solution of the original problem, and the penalty parameters do not need to go to infinity.

Here, the original problem is abstracted into a general nonlinear programming problem form to illustrate the relaxed convex problem of an augmented Lagrange form, as shown below:

$$P_1 \min\ J = \int_0^{\tau_f} 1\ dt = \tau_f$$
$$\text{s.t. } h_i(\hat{x}) = 0, i \in E$$
$$g_i(\hat{x}) \leq 0, i \in I$$
$$s_i(\hat{x}) \in K, i \in C$$

where $\hat{x} = [x, u, t_f]^T$ are augmented optimization variables, all equality constraints are denoted as $h(\hat{x})$, all inequality constraints are expressed as $g(\hat{x})$, and $s(\hat{x})$ represents the second-order cone constraints. And E, I, and C are the corresponding feasible domain sets, respectively.

Then, the slack variables ζ_h, ζ_g, penalty parameters p, and Lagrange multipliers μ, λ are introduced to construct the relaxed problem of augmented Lagrange function form as follows:

$$P_2 \min\ \widetilde{J}(\hat{x}, \zeta_h, \zeta_g) = \tau_f + \lambda^T \zeta_h + p|\zeta_h| + \mu^T \zeta_g + p|\zeta_g|$$
$$\text{s.t. } h_i(\hat{x}) = \zeta_{i,h}, i \in E$$
$$g_i(\hat{x}) \leq \zeta_{i,g}, \zeta_{i,g} \geq 0, i \in I$$
$$s_i(\hat{x}) \in K, i \in C$$

where, for further transformation, the penalty term of the inequality function is not treated as a standard cutoff function because the slack variables ζ_g are nonnegative but as an absolute value term. The similar treatment method can also be found in Ref. [25].

Finally, the problem P_2 is convexified by the first-order Taylor expansion:

$$P_3 \min\ \widetilde{J}(\Delta \hat{x}, \zeta_h^k, \zeta_g^k) = \tau_f + (\lambda^k)^T \zeta_h^k + p^k \left|\zeta_h^k\right| + (\mu^k)^T \zeta_g^k + p^k \left|\zeta_g^k\right|$$
$$\text{s.t. } h_i(\hat{x}^k) + \nabla h_i(\hat{x}^k) \Delta \hat{x} = \zeta_{i,h}^k, i \in E$$
$$g_i(\hat{x}^k) + \nabla g_i(\hat{x}^k) \Delta \hat{x} \leq \zeta_{i,g}^k, \zeta_{i,g}^k \geq 0, i \in I$$
$$s_i(\hat{x}^k + \Delta \hat{x}) \in K, i \in C$$

The objective function of the problem P_3 contains an absolute value term, so it is not a convex second-order cone problem. For this purpose, auxiliary control variables η can be introduced to convert the objective function with absolute values into a combination of the linear objective function and the second-order cone constraints [29]. Specifically, an equivalent unconstrained optimization problem is given below

$$\min\ \sum p|x|,\ p > 0.$$

The minimization problem is equivalent to

$$\min\ \sum p\eta$$
$$\text{s.t. } |x| \leq \eta$$

in which the constraint condition is a second-order cone function, and it is naturally convex.

At this point, a standard convex second-order cone problem P_{SOCP} is represented as:

$$P_{SOCP} \min f_0^T \tilde{x}$$
$$s.t. \quad F\tilde{x} = g_0$$
$$\|M_i\tilde{x} + n_i\|_2 \leq c_i^T \tilde{x} + d_i, \; i = 1, 2, \ldots, (N_C + N_I) \; (*)$$

in which it contains linear objective function, affine equality constraints, and second-order cone constraints. And $\tilde{x} = [\hat{x}, \zeta_h, \zeta_g, \eta]^T$ is the set of the augmented optimization variables, slack variables, and auxiliary control variables. Suppose N_x is the dimension of $\tilde{x}$ after the discretization. The variables N_C and N_I are the number of the second-order cone and inequality constraints, respectively, and let N_E be the number of the equality constraints. The parameters $f_0 \in \mathbb{R}^{N_x \times 1}$, $F \in \mathbb{R}^{(N_C+N_E+N_I) \times N_x}$, $g_0 \in \mathbb{R}^{(N_C+N_E+N_I) \times 1}$, $M_i \in \mathbb{R}^{N_{xi} \times N_x}$, $n_i \in \mathbb{R}^{N_{xi} \times 1}$, $c_i \in \mathbb{R}^{N_x \times 1}$, and $d_i \in \mathbb{R}$ are calculated based on the previous iteration solution $\tilde{x}^{k-1} = [\hat{x}^{k-1}, \zeta_h^{k-1}, \zeta_g^{k-1}, \eta^{k-1}]^T$. It should be noted that the linear inequality constraints can be included in equation (*) because it is a particular case of the second-order cone constraints. And the final convex subproblem P_{SOCP} can be solved iteratively by the advanced interior point method.

For the penalty parameters and Lagrange multipliers in a discrete convex problem P_3, the specific updating methods are as follows:

$$p^k = \rho_p p^{k-1}, \tag{27}$$

$$\lambda_i^k = \lambda_i^{k-1} + 2p^{k-1} h_i(\hat{x}^{k-1}), \tag{28}$$

$$\mu_i^k = \max\left\{\mu_i^{k-1} + 2p^{k-1} g_i(\hat{x}^{k-1}), 0\right\}, \tag{29}$$

where $\rho_p > 1$ is the update multiple of the penalty parameters.

Additionally, to accelerate the convergence of the algorithm, an update approach is designed for the trust region radius in the second-order cone constraints. Let k be the number of iterations and $k_{sc} > 1$ be the number of iterations in which the trust region starts to update. When $k \geq k_{sc}$, if $\tilde{J}(k) \leq \tilde{J}(k-1)$, $\delta_k = b_0 \delta_{k-1}$; otherwise, $\delta_k = b_1 \delta_{k-1}$, where $0 < b_0 < 1 < b_1$ are the trust region contraction factors. Hereon, the trust region is updated in the middle step of the iterations, rather than in the first step. It is to ensure that there is a large manually set scope of trust regions in the early iterations when the system has a certain approximation error. In this way, the algorithm can easily find a feasible optimization direction and perform continuous iterations.

3.3. Solution Procedure

The solution procedure of the AL-CP-ISC to find the solution to the original problem is given. As shown in Figure 1, the convex subproblem can be solved iteratively until predetermined convergence precision is reached.

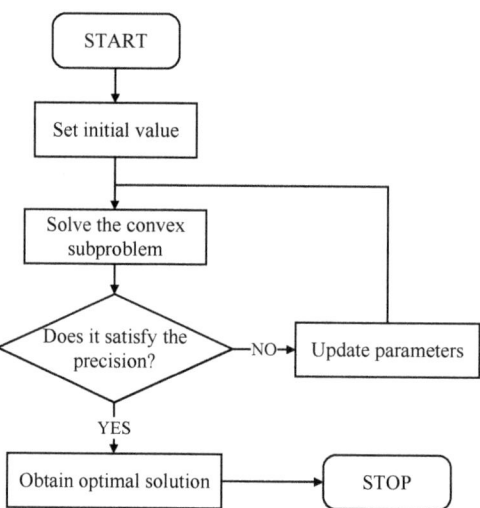

Figure 1. Algorithm solution flow.

A more detailed solution procedure of the AL-CP-ISC algorithm is shown as follows (Algorithm 1):

Algorithm 1. AL-CP-ISC
---
1. Let $k = 0$, set the initial reference trajectory $\hat{x}^0$ by propagating the dynamical Equation (1) with the fixed control variables.
2. Assign initial values to the following parameters: penalty parameters p_0, penalty parameter update multiple $\rho_p > 1$, initial Lagrange multipliers λ_0, μ_0, initial trust region radius δ_0, the iteration number of the trust region starts to update k_{sc}, the trust region contraction factor $0 < b_0 < 1 < b_1$, and the number of discrete points N.
3. $k \geq 1$, solve the convex subproblem P_{SOCP} by the interior point method, and find solution pairs: $\hat{x}^k = [x^k, u^k, t_f^k, \zeta_h^k, \zeta_g^k]^T$.
4. Define the value of constraint violation v:
$$v_k = \sqrt{\sum \zeta_{i,h}^k \cdot \zeta_{i,h}^k + \sum \left| \max\left(\zeta_{i,g}^k, -\mu_i^{k-1}/p_{k-1}\right)\right|} \tag{30}$$
When $v_k \leq \kappa$, go to 6, otherwise, go to 5, where κ is a sufficiently small positive number.
5. Update penalty parameters p_k and Lagrange multipliers λ_k, μ_k.
Then, $k = k + 1$, and go to 3.
6. Obtain the optimal solution of the original problem:
$$\hat{x}^* = [x^*, u^*, t_f^*, \zeta_h^*, \zeta_g^*]^T = [x^k, u^k, t_f^k, \zeta_h^k, \zeta_g^k]^T.$$

Remark 1. *In Step 1, the initial reference trajectory generated by the numerical integration method meets the dynamic constraints, and the change in the bank angle and its rate are smooth. It is beneficial to enable linearization and facilitate the iteration. Nevertheless, notice that the quality of this initial trajectory may affect the convergence effect, such as the number of iterations and convergence accuracy. Therefore, taking into account factors such as the flight time, mission characters, and other constraint conditions, it is necessary to judiciously select the fixed control variable.*

4. Numerical Verification

In this section, numerical simulations are employed to testify the effectiveness and convergence of the AL-CP-ISC algorithm. To compare with the simulation results of the AL-CP-ISC algorithm, CPM and P-CP-ISC algorithms are used to solve the same reentry trajectory optimization problem of the hypersonic vehicle. The CPM is the Chebyshev pseudo-spectral method that transforms the original problem into a general nonlinear programming problem and adopts the sequential quadratic programming algorithm to solve the problem. The difference between P-CP-ISC and AL-CP-ISC lies in different transformation approaches, and the objective function of the P-CP-ISC algorithm adopts the form of a penalty function without Lagrange multipliers.

According to Ref. [30], it is considered that the mass of the reentry vehicle is $m = 104,035$ kg, and the aerodynamic reference area is $S = 391.22$ m^2. In this paper, the fight mission of the vehicle is to plan an optimal trajectory with the minimum flight time under various constraints. Correspondingly, the initial and terminal constraints are: $h_0 = 100$ km, $\theta_0 = 0°$, $\phi_0 = 0°$, $V_0 = 7450$ m/s, $\gamma_0 = -0.5°$, $\psi_0 = 0°$, $\sigma_0 = 1°$, $h_f = 25$ km, $\theta_f = 12°$, $\phi_f = 72°$, $500 \leq V_f \leq 1500$ m/s, $\gamma_f = -10°$, $\psi_f = 90°$. The maximum values of the path constraints are $\dot{Q}_{max} = 1500$ kW/m^2, $q_{max} = 18,000$ N/m^2, and $n_{max} = 2.5g_0$, respectively. In addition, to keep enough control margin, the magnitude of the bank angle and bank angle rate are limited by 80° and 10°/s. The longitude and latitude of the center of the NFZ constraint are $\theta_c = 2°$ and $\phi_c = 50°$, respectively, and the radius is about 222 km. The initial trust region size in Equation (26) is given as:

$$\delta_0 = [\frac{15000}{R_0}, \frac{25\pi}{180}, \frac{25\pi}{180}, \frac{500}{V_0}, \frac{25\pi}{180}, \frac{25\pi}{180}, \frac{25\pi}{180}]^T.$$

In the solution procedure of the AL-CP-ISC algorithm, the termination condition is set to be $v_k \leq 10^{-9}$, and the number of CGL points is 80. In Step 1 of the algorithm procedure, the initial reference trajectory is generated by the numerical integration method, in which the fixed control variable is set to 0.015 °/s, and the estimated terminal time is 1610 s. All optimization algorithms in this section are implemented by MATLAB 2020b on a desktop computer equipped with Intel Core i7-10700K/3.80 GHz CPU. For the convex subproblems P_{SOCP}, the professional software MOSEK (Version 9. 3. 7.) [31] is used to solve them.

First, Figures 2–7 present the trajectories of the vehicle obtained by the three algorithms, and Table 1 shows all the optimal solutions and solve time of each algorithm. The solid red curves are the solution of the AL-CP-ISC, the blue double lines represent the results of the CPM, and the other black dotted lines are the ones of the P-CP-ISC. In Figure 2, the profiles of each iteration of the AL-CP-ISC algorithm are depicted in different colors, from dark blue to warm red, to make the progression of the convergence clearer. The sequence solution converges in the 25th iteration, and the corresponding minimum flight time is 1636.68 s, which is quite close to the optimal solution from the other two algorithms. The trajectory curves almost overlap and are hard to distinguish in late iterations according to the zoom-in view. Moreover, Figures 3–6 show a comparison of results between state variables and control variables, including altitude, velocity, longitude, latitude, flight-path angle, heading angle, bank angle, and its rate. It is obvious that all state and control profiles tend to be relatively consistent, and the initial, terminal, and bounded conditions required by the original problem are satisfied. In particular, it can be seen from Figure 4 that the trajectory of the vehicle successfully evades the no-fly zone. Meantime, the vehicle trajectories obtained by the three algorithms all meet the conditions of path constraints in Figure 7. In Table 1, in addition to the similar optimal solutions, it is worth noting that the solve time of AL-CP-ISC and P-CP-ISC is only 4.20 s and 5.30 s, far less than the solve time required by CPM. Given these analyses, it can be fully explained that the improvement measures of the sequential convexification algorithm do not affect the optimal solution, and the AL-CP-ISC algorithm can vastly reduce the computational cost; that is, the effectiveness and efficiency of the proposed algorithm are verified.

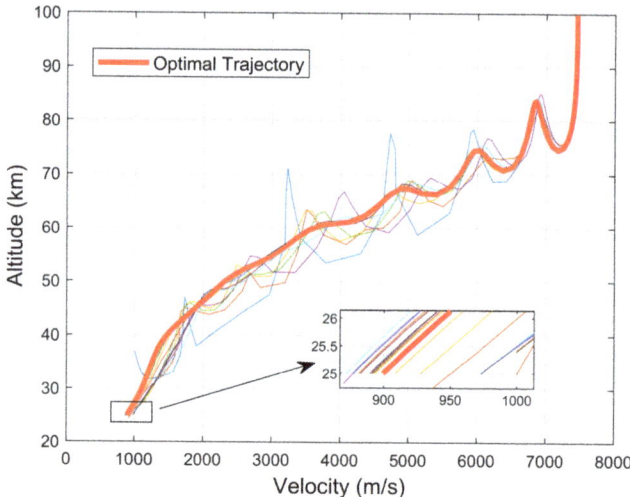

Figure 2. Convergence of the trajectories using AL-CP-ISC.

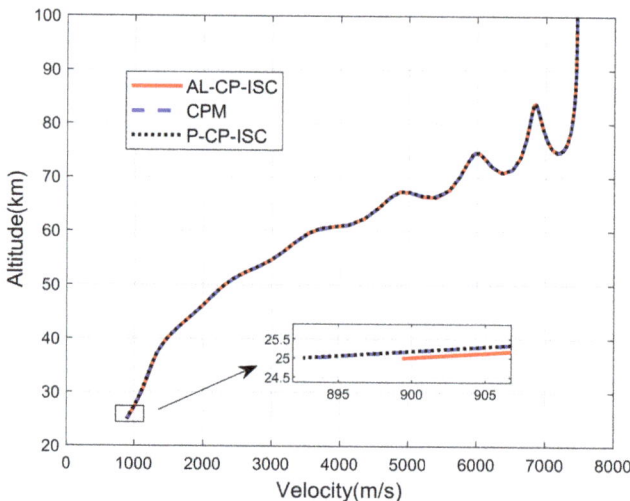

Figure 3. Comparison of altitude–velocity profile.

Then, the state trajectories of the AL-CP-ISC and P-CP-ISC algorithms are almost the same, and some slight differences are seen in the optimal flight time and the solve time. This shows that both algorithms can solve the approximate convex problem well. However, the number of iterations required by both algorithms is quite different according to Figures 8 and 9, and they converge at the 25th and 32th iterations, respectively. Both figures are the value of constraint violations and the terminal flight time in each iteration, respectively. The only difference between the two algorithms is the form of the objective function. In a numerical sense, it indicates that the AL-CP-ISC algorithm with the augmented Lagrange function formal cost function has better convergence properties by comparing the trajectory results of these two algorithms.

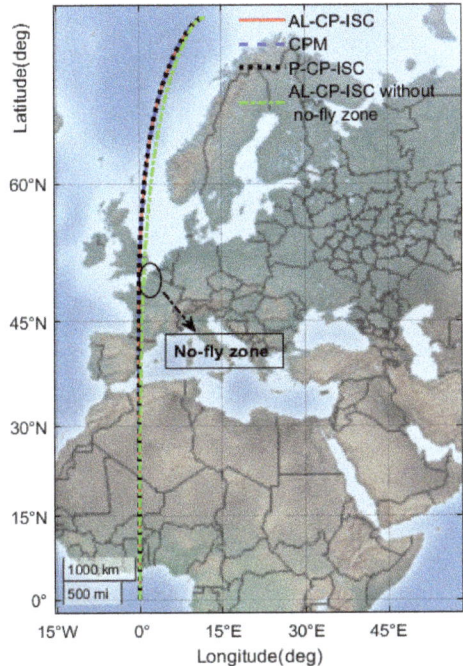

Figure 4. Trajectories profile to avoid the NFZ.

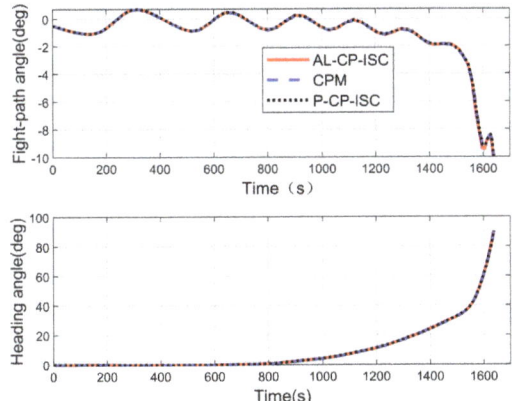

Figure 5. Comparison of fight-path angle and heading angle profile.

Table 1. Comparison of the simulation results.

Algorithm	Flight Time	h_f	θ_f	ϕ_f	V_f	γ_f	ψ_f	Solve Time
CPM	1638.91 s	25 km	12 deg	72 deg	892.71 m/s	−10 deg	90 deg	287.8 s
AL-CP-ISC	1636.68 s	25 km	12 deg	72 deg	899.36 m/s	−10 deg	90 deg	4.20 s
P-CP-ISC	1638.94 s	25 km	12 deg	72 deg	892.63 m/s	−10 deg	90 deg	5.30 s

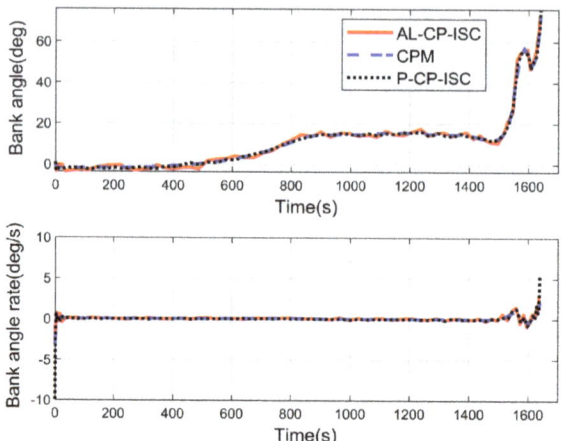

Figure 6. Comparison of bank angle and bank angle rate profile.

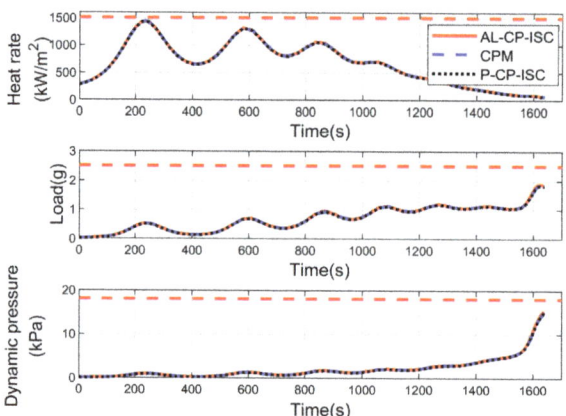

Figure 7. Comparison of the path constraints.

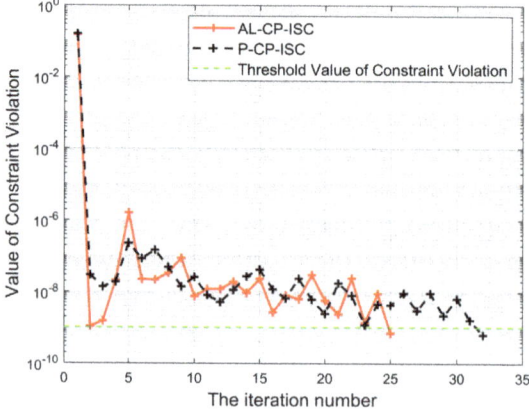

Figure 8. Value of constraints violation for each iteration.

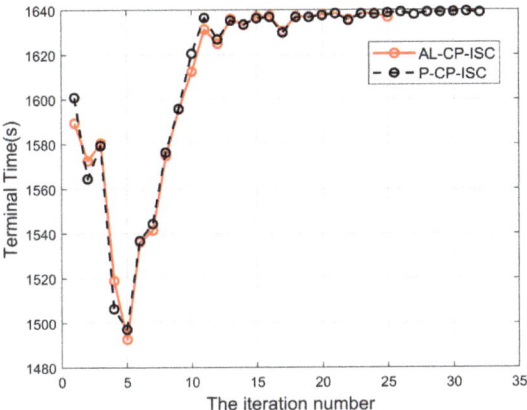

Figure 9. Flight terminal time for each iteration.

5. Conclusions

In this paper, an improved sequential convexification algorithm is proposed to efficiently solve the trajectory planning problem of hypersonic vehicles based on the Chebyshev pseudo-spectral method and the method of multipliers. By employing the Chebyshev pseudo-spectral discretization method, the differential dynamical equations are transformed into the algebraic equation constraint. The original problem is then discretized into a finite-dimensional nonlinear programming problem. Next, through linearization and relaxation techniques, the slack variables, penalty parameters, and multipliers are introduced to transform the discrete problem into a series of relaxed convex subproblems in the form of the augmented Lagrange function, which are iteratively and efficiently solved by the MOSEK solver until the expected convergence precision is satisfied. The numerical simulation results verify the effectiveness, efficiency, and convergence performance of the proposed AL-CP-ISC algorithm. In the future, the proposed algorithm will be competitive for use in realizing onboard optimization due to its excellent optimization efficiency after code optimization.

Author Contributions: Conceptualization, Y.L. and J.W.; investigation, J.Z.; methodology, Y.L., H.C. and J.W.; validation, Y.L. and H.C.; resources, H.L.; writing—original draft preparation, Y.L.; writing—review and editing, J.W. and J.Z.; funding acquisition, H.L. All authors have read and agreed to the published version of the manuscript.

Funding: The research presented in this paper is supported by the National Natural Science Foundation of China (grant number 62103452) and Shenzhen Science and Technology Program (grant number ZDSYS20210623091808026).

Data Availability Statement: The data presented in this study are available on request from the corresponding author.

Acknowledgments: The authors acknowledge Jinbo Wang from the School of Systems Science and Engineering of Sun Yat-sen University for his support and guidance.

Conflicts of Interest: The authors declare no conflicts of interest.

Appendix A

All elements of the matrix A in Equation (17) are given as follows.

$$A = (t_0 - t_f)\frac{\partial f_0(x,t)}{\partial x} = (t_0 - t_f)\begin{bmatrix} 0 & 0 & 0 & a_{14} & a_{15} & 0 & 0 \\ a_{21} & 0 & a_{23} & a_{24} & a_{25} & a_{26} & 0 \\ a_{31} & 0 & 0 & a_{34} & a_{35} & a_{36} & 0 \\ a_{41} & 0 & 0 & a_{44} & a_{45} & 0 & 0 \\ a_{51} & 0 & 0 & a_{54} & a_{55} & 0 & a_{57} \\ a_{61} & 0 & a_{63} & a_{64} & a_{65} & a_{66} & a_{67} \\ 0 & 0 & 0 & 0 & 0 & 0 & 0 \end{bmatrix}$$

where

$$a_{14} = \sin\gamma, a_{15} = v\cos\gamma$$

$$a_{21} = -\frac{v\cos\gamma\sin\psi}{r^2\cos\phi}, a_{23} = \frac{v\cos\gamma\sin\psi\sin\phi}{r\cos^2\phi}, a_{24} = \frac{\cos\gamma\sin\psi}{r\cos\phi},$$

$$a_{25} = -\frac{v\sin\gamma\sin\psi}{r\cos\phi}, a_{26} = \frac{v\cos\gamma\cos\psi}{r\cos\phi}$$

$$a_{31} = -\frac{v\cos\gamma\cos\psi}{r^2}, a_{34} = \frac{\cos\gamma\cos\psi}{r}, a_{35} = -\frac{v\sin\gamma\cos\psi}{r}, a_{36} = -\frac{v\cos\gamma\sin\psi}{r}$$

$$a_{41} = -D_r + \frac{2\sin\gamma}{r^3}, a_{44} = -D_v, a_{45} = -\frac{\cos\gamma}{r^2}$$

$$a_{51} = \frac{L_r\cos\sigma}{v} - \frac{v\cos\sigma}{r^2} + \frac{2\cos\gamma}{r^3 v}, a_{54} = \frac{L_v\cos\sigma}{v} - \frac{L\cos\sigma}{v^2} + \frac{\cos\gamma}{r} + \frac{\cos\gamma}{v^2 r^2}$$

$$a_{55} = -\frac{v\sin\gamma}{r} + \frac{\sin\gamma}{vr^2}, a_{57} = -\frac{L\sin\sigma}{v}$$

$$a_{61} = \frac{L_r\sin\sigma}{v\cos\gamma} - \frac{v\cos\gamma\sin\psi\tan\phi}{r^2} + \frac{2\cos\gamma}{vr^3}, a_{63} = \frac{v\cos\gamma\sin\psi}{r\cos^2\phi}$$

$$a_{64} = \frac{L_v\sin\sigma}{v\cos\gamma} + \frac{\cos\gamma\sin\psi\tan\phi}{r}, a_{65} = \frac{L\sin\sigma\sin\gamma}{v\cos^2\gamma} - \frac{v\sin\gamma\sin\psi\tan\phi}{r}$$

$$a_{66} = \frac{v\cos\gamma\cos\psi\tan\phi}{r}, a_{67} = \frac{L\cos\sigma}{v\cos\gamma}$$

$$D_r = \frac{\partial D}{\partial r} = -\frac{R_0^2 \rho v^2 S C_D}{2mh_s} = -\frac{R_0}{h_s}D, D_v = \frac{\partial D}{\partial v} = \frac{R_0 \rho v S C_D}{m}$$

$$L_r = -\frac{R_0}{h_s}L, L_v = \frac{R_0 \rho v S C_L}{m}$$

References

1. Betts, J.T. Survey of numerical methods for trajectory optimization. *J. Guid. Control Dyn.* **1998**, *21*, 193–207. [CrossRef]
2. Chai, R.; Savvaris, A.; Tsourdos, A.; Chai, S.; Xia, Y. A review of optimization techniques in spacecraft flight trajectory design. *Prog. Aerosp. Sci.* **2019**, *109*, 100543. [CrossRef]
3. Zhang, Y.; Zhang, R.; Li, H. Mixed-integer trajectory optimization with no-fly zone constraints for a hypersonic vehicle. *Acta Astronaut.* **2023**, *207*, 331–339. [CrossRef]
4. Sagliano, M.; Mooij, E. Optimal drag-energy entry guidance via pseudospectral convex optimization. *Aerosp. Sci. Technol.* **2021**, *117*, 106946. [CrossRef]
5. Wang, J.; Liang, H.; Qi, Z.; Ye, D. Mapped Chebyshev pseudospectral methods for optimal trajectory planning of differentially flat hypersonic vehicle systems. *Aerosp. Sci. Technol.* **2019**, *89*, 420–430. [CrossRef]
6. Mao, Y.; Zhang, D.; Wang, L. Reentry trajectory optimization for hypersonic vehicle based on improved Gauss pseudospectral method. *Soft Comput.* **2017**, *21*, 4583–4592. [CrossRef]
7. Zhou, H.; Wang, X.; Cui, N. Glide trajectory optimization for hypersonic vehicles via dynamic pressure control. *Acta Astronaut.* **2019**, *164*, 376–386. [CrossRef]
8. Wang, J.; Cui, N.; Wei, C. Rapid trajectory optimization for hypersonic entry using a pseudospectral-convex algorithm. *Proc. Inst. Mech. Eng. Part G J. Aerosp. Eng.* **2019**, *233*, 5227–5238. [CrossRef]

9. Wang, Z.; Grant, M.J. Constrained trajectory optimization for planetary entry via sequential convex programming. *J. Guid. Control. Dyn.* **2017**, *40*, 2603–2615. [CrossRef]
10. Ma, Y.; Pan, B.; Hao, C.; Tang, S. Improved sequential convex programming using modified Chebyshev–Picard iteration for ascent trajectory optimization. *Aerosp. Sci. Technol.* **2022**, *120*, 107234. [CrossRef]
11. Boris, B.; Alessandro, Z.; Guido, C.; Pizzurro, S.; Cavallini, E. Convex optimization of launch vehicle ascent trajectory with heat-flux and splash-down constraints. *J. Spacecr. Rocket.* **2022**, *59*, 900–915.
12. Acikmese, B.; Ploen, S.R. Convex programming approach to powered descent guidance for mars landing. *J. Guid. Control. Dyn.* **2007**, *30*, 1353–1366. [CrossRef]
13. Blackmore, L.; Açikmeşe, B.; Scharf, D.P. Minimum-landing-error powered-descent guidance for Mars landing using convex optimization. *J. Guid. Control Dyn.* **2010**, *33*, 1161–1171. [CrossRef]
14. Wang, Z.; Grant, M.J. Optimization of minimum-time low-thrust transfers using convex programming. *J. Spacecr. Rocket.* **2018**, *55*, 586–598. [CrossRef]
15. Lu, P.; Liu, X. Autonomous trajectory planning for rendezvous and proximity operations by conic optimization. *J. Guid. Control Dyn.* **2013**, *36*, 375–389. [CrossRef]
16. Liu, X.; Lu, P. Robust trajectory optimization for highly constrained rendezvous and proximity operations. In Proceedings of the AIAA Guidance Navigation and Control (GNC) Conference, Boston, MA, USA, 15 August 2013; p. 4720.
17. Wang, L.; Ye, D.; Xiao, Y.; Kong, X. Trajectory planning for satellite cluster reconfigurations with sequential convex programming method. *Aerosp. Sci. Technol.* **2023**, *136*, 108216. [CrossRef]
18. Wang, J.; Cui, N.; Wei, C. Rapid trajectory optimization for hypersonic entry using convex optimization and pseudospectral method. *Aircr. Eng. Aerosp. Technol.* **2019**, *91*, 669–679. [CrossRef]
19. Liu, X.; Shen, Z.; Lu, P. Entry trajectory optimization by second-order cone programming. *J. Guid. Control Dyn.* **2016**, *39*, 227–241. [CrossRef]
20. Pei, P.; Fan, S.; Wang, W.; Lin, D. Online reentry trajectory optimization using modified sequential convex programming for hypersonic vehicle. *IEEE Access* **2021**, *9*, 23511–23525. [CrossRef]
21. Fahroo, F.; Ross, I.M. Direct trajectory optimization by a Chebyshev pseudospectral method. *J. Guid. Control Dyn.* **2002**, *25*, 160–166. [CrossRef]
22. Mittal, A.K.; Balyan, L.K. An improved pseudospectral approximation of coupled nonlinear partial differential equations. *Int. J. Comput. Sci. Math.* **2022**, *15*, 155–167. [CrossRef]
23. Wang, Z.; Lu, Y. Improved sequential convex programming algorithms for entry trajectory optimization. *J. Spacecr. Rocket.* **2020**, *57*, 1373–1386. [CrossRef]
24. Wang, Z. Optimal trajectories and normal load analysis of hypersonic glide vehicles via convex optimization. *Aerosp. Sci. Technol.* **2019**, *87*, 357–368. [CrossRef]
25. Kanzow, C.; Nagel, C.; Kato, H.; Fukushima, M. Successive linearization methods for nonlinear semidefinite programs. *Comput. Optim. Appl.* **2005**, *31*, 251–273. [CrossRef]
26. Nocedal, J.; Wright, S.J. *Numerical Optimization*; Springer: New York, NY, USA, 1999.
27. Lu, Z.; Sun, Z.; Zhou, Z. Penalty and augmented Lagrangian methods for constrained DC programming. *Math. Oper. Res.* **2022**, *47*, 2260–2285. [CrossRef]
28. Liu, X.-W.; Dai, Y.-H.; Huang, Y.-K.; Sun, J. A novel augmented Lagrangian method of multipliers for optimization with general inequality constraints. *Math. Comput.* **2023**, *92*, 1301–1330. [CrossRef]
29. Liu, X.; Lu, P. Solving nonconvex optimal control problems by convex optimization. *J. Guid. Control Dyn.* **2014**, *37*, 750–765. [CrossRef]
30. Stanley, D.O.; Engelund, W.C.; Lepsch, R.A.; McMillin, M.; Wurster, K.E.; Powell, R.W.; Guinta, T.; Unal, R. Rocket-powered single-stage vehicle configuration selection and design. *J. Spacecr. Rocket.* **1994**, *31*, 792–798. [CrossRef]
31. Andersen, E.D.; Roos, C.; Terlaky, T. On implementing a primal-dual interior-point method for conic quadratic optimization. *Math. Program.* **2003**, *95*, 249–277. [CrossRef]

Disclaimer/Publisher's Note: The statements, opinions and data contained in all publications are solely those of the individual author(s) and contributor(s) and not of MDPI and/or the editor(s). MDPI and/or the editor(s) disclaim responsibility for any injury to people or property resulting from any ideas, methods, instructions or products referred to in the content.